Rick Snoeyink

Rick Snoeyink

Advanced Interactive Video Design

New Techniques and Applications

by Nicholas V. Iuppa
with Karl Anderson

Knowledge Industry Publications, Inc.
White Plains, NY

Video Bookshelf

Advanced Interactive Video Design:
New Techniques and Applications

Library of Congress Cataloging in Publication Data
Iuppa, Nicholas V.
    Advanced interactive video design.

      (the Video bookshelf)
    Bibliography: p.
    Includes index.
    1. Interactive video—Design and construction.
I. Anderson, Karl. II. Title. III. Series.
TK6643.I97   1987   384.55   87-22659
ISBN 0-86729-170-2

Printed and bound in the United States of America

10   9   8   7   6   5   4   3   2

Table of Contents

List of Tables and Figures

ACKNOWLEDGMENTS

I wish to dedicate this book to all the managers who supported, subsidized and otherwise encouraged our work in the development of interactive video. Of these, I would especially like to thank Jack Evard, Dave Mebane and Leonard Linden at Bank of America, Marika Ruumet at Hewlett-Packard Co. and Gene Lipkin at ByVideo, Inc.

Author's Note

The views in this book are solely my own and not those of Apple Computer, Inc. This book was written prior to my employment at Apple and Apple has not reviewed the book, and in no way authorizes any of my statements.

Nicholas V. Iuppa

1 Introduction

Interactive compact disc systems like CD-I (Compact Disc-Interactive), CDV (Compact Disc-Video), CVD (Compact Videodisc) and DVI (Digital Video Interactive) are joining laser videodisc products in today's marketplace. Unfortunately, the appearance of these different and new interactive systems is prompting people to raise the cry, "Wait until they standardize! Let's not bother to learn anything about interactive video until the industry settles down and decides on the one format that will be the standard forevermore." There are quite a number of reasons why such an attitude is wrong.

Interactive discs that have grown out of compact disc technology and that have piggybacked on the recognized high quality of that successful audio format, may prove to be even more successful and may, in some instances, replace the current 12-inch laser videodisc format. But whether the industry will standardize on the compact disc or the 12-inch disc, on CD or LDP technology, or whether everyone will suddenly turn to a new *tape* format, has little to do with the study we are about to undertake in this book. We are here to talk about interactive video program *design*.

Interactive video professionals generally agree that the design of interactive video programs is one of the most difficult things to understand. It is the one element that transfers across all the different interactive media and it is very difficult to do well. Whether your next interactive video program will be on disc, CD, videotape or a new medium that has not been heard of yet, most of the design principles and organizational concepts that are presented in this book will work for you.

It could and probably should be argued that the unsuccessful development and sale of interactive disc systems to this date can be attributed to the fact that there have been so few examples of successful interactive video programs. The fact is, when we all started designing interactive video programs in the early 1980s we did not know how to put the programs together effectively; we did not know the best organization to use; we had not found the best flow to give to an interactive video

1

story; and we did not know the proper logic that would make programs immediately usable by the general public.

Nothing was obvious and it still isn't. There have now been almost ten years of interactive design and experimentation going on, but very little of it has lead to smashing sales success. What we have obtained, however, is a growing body of knowledge that constitutes the science and art of interactive design.

There were over 20 years between the time that movies were invented and the moment when D. W. Griffith called for the first close-up movie shot. Yet, many current techniques in movie-making would be unthinkable without close-ups.

Today we are at the same point in the evaluation of interactive video. We are still inventing, identifying and cataloging the tools of our trade. The indispensible element of interactive video may not yet have come to the fore.

This book is our best effort to define the major tools in the interactive video toolbox, so they can be made accessible to everyone. Many of the techniques that we will describe are advanced, others are just little known, but they need to be added to the stock in trade of all interactive video designers. Certainly the list is not complete. This is an evolving technology, one that has the potential to become an art form in its own right. Let's work so that the technology can grow and become *all* that it can be.

THE SEARCH FOR THE BREAKTHROUGH APPLICATION

If you look back over the years since the interactive video medium first emerged, you will see that a number of breakthrough applications appeared to be on the horizon–applications that would make interactive video as popular as TV, toothpaste, or light bulbs. Somehow they never happened.

Entertainment Video

At first, entertainment video was a likely candidate to capture the public's imagination. VCRs were just becoming popular, but discs offered better picture, better sound, better special effects, a lower per-program price and an undefined quality called *interactivity*. Who would have guessed that people would pass over all of these incredible videodisc features for the one feature VCR offered that discs did not: the ability to record.

Actually, everyone should have expected this reaction. But at the time (around 1980) no one thought that people would prefer to *rent* video programs rather than buy them. Maybe that should have been obvious, too.

Perhaps entertainment discs would have been more successful if interactive entertainment programs were available–interactive entertainment disc programs that were worth using. But the consumer machines of the time were only capable of limited interactivity and no one really knew how to produce interactive discs even if there had been machines that were capable of playing them.

All in all, entertainment discs just were not a viable alternative to video cassettes, nor, we think, did the manufacturers want them to be. After all, millions, if not billions were to be made from the sale of VCRs so, why confuse the issue?

Interactive Video Training

Corporate training presented, perhaps, a better means of becoming a breakthrough application, especially since there were a great many instructional designers whose experience with the design and development of programmed instruction gave them insight into the concept of interactivity. Although the real problem in the case of training appeared to be bad timing, it may just have been bad luck on the part of the disc manufacturers that caused their efforts to fail. In 1982, many of their potential customers were several years into the eight year amortization of 3/4-inch videotape systems. These corporate training executives could not have sold their management new training hardware no matter how wonderful it was.

Interactive Games

By 1984, many people expected interactive videodisc games to be the breakthrough application they were waiting for. When children tired of the traditional video games, it was thought that videodisc games would replace them and offer the game companies another chance. That actually started to happen. *Dragon's Lair,* manufactured by Cinematronics, Inc. was the first and perhaps the best of the videodisc arcade games, and it spawned many imitators. The only problem was that the new game makers forgot that videodisc games had to be fun to play or they would fail. The mad scramble to get videodisc games onto the market after the initial success of *Dragon's Lair* resulted in the production of all kinds of games that, in one way or another used discs. Many of these games, however, were not exciting enough and did not fit the definition of those who knew what interactive disc games could be, and they failed.

Point-of-Sale Disc Systems

In late 1984, point-of-sale (POS) disc systems captured the attention of the disc prognosticators. Here was an application that fit right into the mold of the automated teller machine. People could go up to interactive videodisc information centers and gain advice about travel, shopping, communication and entertainment. These centers offered 24-hour service and a variety of features. There were only two problems with POS systems: No one seemed to be able to show how the systems could make money without actually selling products and, when the POS system manufacturers tried selling products, they ran into a new and unforessen problem—people are quite selective about who they buy from and they are most resistant to buying from machines.

Level 3 Interactive Videodisc Programs

So, here we are again, searching for the breakthrough application for interactive discs, and wondering what it might be. There are many possibilities; some are new and some are extensions of the "hoped-for" breakthrough applications of the past. What all of these new applications have in common is that they tap into computer technology to turn the disc player into a computer peripheral by producing Level 3 interactive videodisc programs. Much of this book will make the point that this is the best of all directions to take now; the one with the most promise.

SOURCE MATERIAL FOR THIS BOOK

This book has grown out of years of experience with the use of video and other media for training, educational, informational and POS purposes. Some of the concepts presented go back to 1966 when Stanford University's Department of Communication produced a series of teacher training films on common classroom problems. The dramatizations led to choices in which the teacher was asked what to do, and the film then presented the consequences of each choice. This is called consequence remediation and is a popular interactive video design technique today.

Maybe the concepts behind this book go back even further—to 1882 and Frank Stockton's short story "The Lady or the Tiger" in which a young nobleman is driven by his captors to choose between two doors, one of the doors has a fierce tiger behind it, the other, a beautiful princess. It is left up to the reader's imagination to determine the outcome. Interactive video is based on the same basic principle- the users' choices effect the outcome.

Regardless of how long the concept of interactivity has been a part of the collective unconscious of the human species, it seems that many people cannot grasp the possibilities of interactivity until they see it in action. Because of this, this book will continually cite actual projects. We will relate our experiences gained through interactive video training programs produced for banks, customer service centers, sales personnel, managers and technicians at a variety of companies. We will also discuss POS and information dissemination systems that we have built for video shopping, consumer product and travel companies. Perhaps most important, we will give concrete examples of uses of interactive video as a peripheral to a personal computer system. These examples, of course, are based on work done for computer hardware and software manufacturers.

OBJECTIVES

We have written this book to help you understand the capabilities of interactive video and to know where to get started in designing creative and useful programs yourself. Upon completion of this book, you should be able to do the following:

- list the most promising applications of interactive video in computer training, education, entertainment, data storage, information dissemination and POS
- recognize the benefits of interactive video when used for training on the use of computer software
- distinguish between interactive video used for "Help Systems" and that used purely for training
- plan the conversion of existing linear video material into interactive programming
- recognize the best methods for organizing educational material into interactive programming
- understand the capabilities of the most promising interactive entertainment applications
- understand the possible uses of interactive video databases
- recognize the most important organizational elements required for successful POS and information dissemination programs
- know where to start to put together the hardware to support Level 3 interactive disc programs
- schedule and budget for the development of an interactive video program or system
- know where to get help in your work with interactive video

Throughout all of this, it is critical to remember that producing interactive video is not like mixing cement in which any approximation of the right formula will do. Making interactive video is more like building an airplane in which everything must be just right or the plane will not get off the ground—and, if it is forced off the ground it will probably crash.

We think it is exciting to be present at the dawn of a technology; to be among those who shape the directions it will take. So, if you're into interactive video, stay with it. If you're not, join us. It's going to be a lot of fun.

2 An Interactive Video Help System

If you use a computer, how did you learn to use it? Did you attend a seminar or did someone show you how to use the computer through one-on-one training on the job? We bet you didn't learn to use it by reading the manual either. Computer manuals are so detailed that they are difficult to finish and they are often so badly written that you cannot learn from them even if you take the time to read them. There must be a better way.

One possibility is to use the "Help System" on the computer which you get when you push the HELP button. We suspect that many people who learn computer software programs do so by going directly to the program and accessing the help mode. But, the help mode is often more poorly written than the manual, and it is often organized haphazardly. There has to be a better way than this, too. And there is. . . .

It's a "Video Help System" that uses the power and effectiveness of interactive video to support the computer software—to offer the computer user, in effect, the perfect learning tool to solve one of the major learning problems of the twentieth century—how to use a given piece of computer software. Moreover, because the system is created by professional instructional designers and written by professional communicators, it works well.

ADVANTAGES OF A VIDEODISC HELP SYSTEM

Seldom have the economics for the introduction of a system been so right. Half of the hardware is already out there—the computers, and almost all of these computers know how to talk to a disc player. Computer users ordinarily spend anywhere from $500 to $3000 for peripheral equipment to help them run their systems, and they are continually looking for ways to maximize their investments. Although they may have spent a great deal of money on computer software, many users have not maximized their investments because they probably will not use the software to its fullest capacity.

Think about it. Think about a word processing program that you have. You can type with it and you can correct with it but, do you know how to underline a word? Do you know how to insert a paragraph? Do you know how to move pages around within a given document? Do you know how to transfer paragraphs or pages to another document? You may be able to do some of these things, but how many of the capabilities of your word processing system are you really using? If you are not using all of them you are most likely wasting your investment in the program.

Most computer users tell us that it would be worth a great deal to them if they could put software programs into their computers and use them right away. A Video Help System could make this happen.

What Is Needed

What would it take for a Video Help System to come into existence? The computer systems are already in place, and the need is certainly there. What's missing is the disc player and maybe the video monitor. Computer users are people who have been conditioned to expand their systems with more and more peripherals. Not only are they used to expansion, they look forward to it. So, lack of hardware may not be as big a hurdle as it has been to other applications of interactive videodisc technology.

Another roadblock to this kind of system is the lack of a prototype system—almost no one has yet built a successful Videodisc Help System. It is certain, however, that such a system would have to be astoundingly good if it were going to lead anyone to buy the necessary hardware, and it would have to be available for many different types of software, not just for one or two programs.

This chapter explains how to build that type of learning system program. We think it is a program that could be the breakthrough application that would open the door for all the other Level 3 applications of interactive video. Once we have reviewed our Video Help System we will go on and explore other applications of interactive video in their most advanced forms.

DEFINING THE INTERACTIVE VIDEO HELP SYSTEM

To fully understand what an interactive Video Help System is, you have to understand the important distinction between job aids and instruction. A job aid is a tool that tells you how to do a job while you're doing it. Instruction is a mechanism that teaches you how to do a job from memory. In other words, training attempts to plant instructions in your head so you will be able to do a job without any help at all. Job aids are designed to be a source of help on the job.

The classic example of a job aid is the cookbook. It contains recipes that you probably cannot remember but which you can look up when you need them. In the same way, you may not be able to remember every single possible function of a word processing system, but with the right kind of job aid, it's easy to have them readily available. If you use job aids (or help systems as we would like to call them) frequently enough, and if they are constructed in the right way, eventually, they will help you learn to do things from memory. It seems clear that, given the computer user's penchant for diving in and using various systems before attending a seminar, or reading a manual, the job aid approach may be one of the very best ways to make the program immediately beneficial and the user immediately productive.

Most computer programs have help systems built into them, but many of these systems are poorly conceived and even the best don't offer the friendliness needed to break the ice and get the user into the program as well as some other media systems can. For these reasons we would like to present our recommended organization for what we feel is the ultimate help system—a help system that uses interactive video as its medium of delivery. The organization is designed to capitalize on the things that video does best, foremost among these being the ability to use clear and understandable communication. It should be noted, however, that this organization would do wonders for computer help systems even if it were applied to them by itself—as an organization without adding video.

MENU DRIVEN VERSUS CONTEXT SENSITIVE

One of the key concerns of help system designers, video or otherwise, is whether the help frames that come to the screen when "Help" is requested are determined by the context in which the request is made or by a menu that comes to the screen before the first real help frame appears.

In the ideal situation, the help system would be a combination of both context and menu drive. In the real world, however, a context sensitive help system cannot be created unless there is a way for the interactive video system to know where you are in the program. To do this, you have to get into the inner workings of the software program you are helping. To get into the inner workings you have to know the company (or person) that designed and holds the copyright (and source code) for the program.

For example, you are using a word processing program and trying to print out a document and you ask for "Help." At this point you should generally get help on how to print out. If the interactive video is not tied into the program it will not know what you are doing. It will not be sensitive to the context of your request and the best it will be able to do is offer a menu of help choices among which "How To

Print" will be available. If you (as a program designer) are fortunate enough to have the ability to gain access to the inner workings of the program and can put in the right hooks to get the user to the appropriate help sequence, then you will understand why context sensitive video helps are preferred.

We did say that the best systems combine context sensitive helps with menu driven helps. These systems enable a user who has asked for help and received some advice to move on and find other possible help materials. Perhaps the best way to begin creating a combined menu/context help system is to create a menu of all the possible help frames. This would allow the eventual user to go to one point to review all the possible help choices. Once this menu has been created, you can go back and assign each of the help frames to a particular function of the program (or to particular error routines).

BEGINNER'S HELP

Most good computer programs operate in levels. That is, people can use parts of the program before they learn to use all the special features that have been created for it. Word processing is again an excellent example of a program that operates in levels. The simplest word processing program operates as a typewriter with a few correction features. One good way to teach the use of such a system is to begin with the basic functions before going on to some of the more sophisticated editing features.

An effective Video Help System, therefore, would offer the beginner an opportunity to access a special help menu. Such a menu would offer help with the indispensable steps that are needed to do basic functions. Figure 2.1 lists nine steps that might be offered in a Beginner's Help Menu for a word processing program. In this figure, if number (1) is selected, the user will proceed through all of the following steps (2 through 9) to gain an overview of the word processing activities. If an individual number (2 through 9) is selected, the appropriate help function and that function only would come up on the screen.

In this example, we have deliberately left out activities such as inserting artwork, moving paragraphs and shuffling pages. These are second-level activities that should be saved for a second-level presentation. We suggest that second-level presentations only be created if the appropriate general menus are also created.

SUMMARY FRAMES

One of the assumptions you have to make when preparing an interactive Video Help System is that it will be used by the same person more than once. Since a help system is not a training device as much as it is a reference tool, people don't

Figure 2.1: Beginner's Help Menu

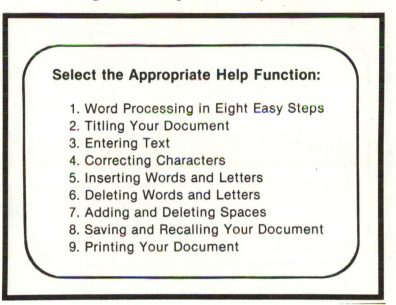

Select the Appropriate Help Function:

1. Word Processing in Eight Easy Steps
2. Titling Your Document
3. Entering Text
4. Correcting Characters
5. Inserting Words and Letters
6. Deleting Words and Letters
7. Adding and Deleting Spaces
8. Saving and Recalling Your Document
9. Printing Your Document

memorize what it says. They usually go back to it again and again. For this reason it would be desirable to accommodate those who understand most of the information and only need a refresher on key parts that are not easy to remember. In the computer world there is no better example of "key parts that are not easy to remember" than computer commands. There are so many commands in each computer program and the commands are so seldom obvious, that it is very easy to confuse or forget them.

Program designers will tell you that long after a program user learns the procedures for operating a program, he or she will continue to forget some of the commands. It is a good idea for the sake of these repeat help users to begin any help section, not with a detailed explanation, but with a simple overview of the appropriate commands.

Figure 2.2 offers a typical frame that might be used to introduce viewers to a video help section. For the purposes of this example, we have changed our subject from word processing to spreadsheet calculation, an operation full of commands. The commands in Figure 2.2 perform basic math calculations.

The video help frame depicted in Figure 2.2 is called a summary frame because it summarizes the material presented in the video explanation. A person who had used this spreadsheet program before, but who could not remember the command for multiplying (an asterisk used as a times sign), would need to go no further than this frame to find out what to do. Someone entering the material for the first time, however, might need much more information.

Figure 2.2: Spreadsheet Video Help Frame

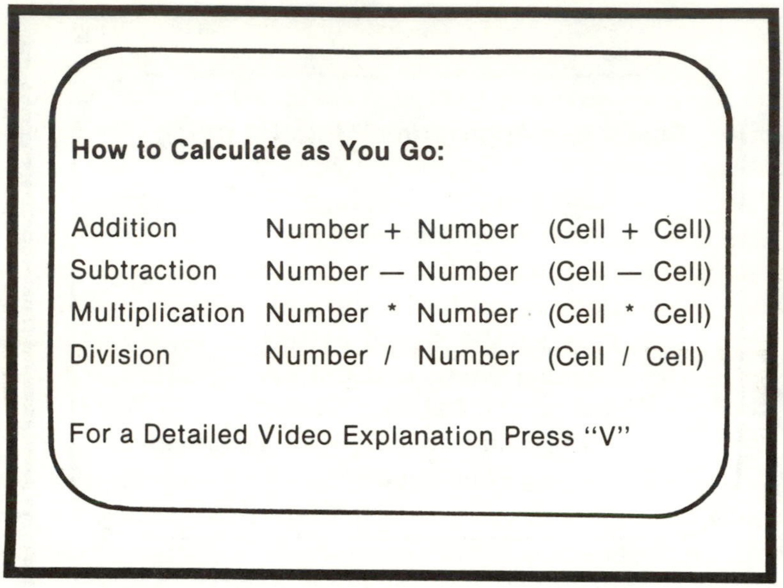

How to Calculate as You Go:
Addition Number + Number (Cell + Cell)
Subtraction Number — Number (Cell — Cell)
Multiplication Number * Number (Cell * Cell)
Division Number / Number (Cell / Cell)
For a Detailed Video Explanation Press "V"

A final note before we explain how to present that video explanation; since the summary frame highlights the entire section presented in the video explanation, we think it is best to repeat the summary frame at the end of the detailed explanation as well as before. Figure 2.3 is a flowchart of the organization of a Video Help System that could be used for a spreadsheet program.

Figure 2.3: Organization of a Possible Spreadsheet Video Help System

Help Menu
Calculations
"How to Calculate Summary Frame"
Detailed Video Explanation

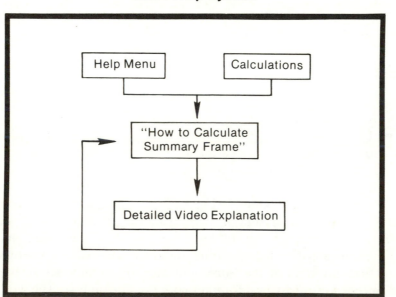

MULTIPLE IMAGE EXPLANATIONS AND OTHER PRODUCTION TIPS

There are quite a number of techniques that can be used to make the detailed explanation as effective as possible. For example, a close-up of the computer screen can be used to show the user exactly how a command or input should look as it is entered. A close-up can also illustrate a typical effect of the command to help the viewer understand exactly what the command does.

Video character generation and other video graphics systems can provide a strong visual support for any explanation. One good way to implement these devices is to use an electric paintbox or other video graphic system to create a representation of the actual computer keys to enhance the realism of the graphics. This is especially helpful when working with special function keys that are hard to describe or find.

The screen image of a person talking to users and showing them how to enter certain data into the computer can also be a valuable component of a Video Help System. Seeing the person who is speaking does help break up an endless stream of graphics, and helps "humanize" the message.

As good as each of these devices are by themselves, however, they are even more powerful when combined into a composite image; a three- or four-way split screen that combines an image of the computer screen, a graphic of the data to be entered and a live image of the person sitting at the terminal talking to the viewer. Moreover, if these scenes are staged and shot properly no digital effects will be needed and the cost will be less than one might imagine.

The three-way split screen described in Figure 2.4 works quite well when it is used to demonstrate command entry. The command is shown three ways. This triple redundancy virtually guarantees that the viewer will get the message.

Developing Detailed Explanations   *not really meant for teachers – would need to make videodisc – $*

It is important to keep the wording of the help narratives clear and to the point. A conversational tone seems to work best although an occasional aside by the narrator may also be well received. Expressions that we've used with success include:

> There is one little trick to this.
>
> or
>
> Believe me. This is one technique you could
> never figure out by trial and error.

We have also discovered that the type of person who delivers the message is also important. Representatives of the target audience are usually the safest bet. In the ideal world the viewer might be able to choose the person he or she would prefer to

Figure 2.4: Three-Way Split Screen

have deliver the instructional message. If there is a social component to on-the-job computer training, having the right Mr. or Ms. "Believable" on video might offer a pleasant and perhaps less frustrating alternative to the process.

In a discussion, to follow, we will be talking about two-screen systems versus one-screen systems. At this point, however, we might just say that in an interactive Video Help System, a two-screen system that offers a video monitor and a separate computer CRT does offer the user the opportunity to hold a help frame on the screen while he or she continues to work with the computer program. This could prove to be very helpful for complex applications.

IDEAL USES OF VIDEO HELP SYSTEMS

Instructions providing information on how to use a computer may very well be the first and best of the interactive Video Help System uses. We have even imagined a *Computer Users Video Handbook* that might serve as the pinnacle of computer instruction. Other obvious uses include word processing help systems and spreadsheet helps, not to mention help systems for filing programs, budgeting programs, graphics programs, mail programs and even specialized applications such as those designed by companies to handle their unique needs (inventories, ordering, billing, etc.). It has also been suggested that it would be useful to have "background helps" that would give "support knowledge" on a given subject. To the "help system" on desktop publishing, for example, you might add a help system on page layout and design.

3 Interactive Video for Training

In its strictest sense, training refers to programs that teach people how to do things from memory. We all know the most common methods for training: classroom instruction, tutoring and self-instruction. Most of us are also well aware of the various instructional media and the roles that film, video, overhead slides and hand-out sheets play in a typical learning experience.

What is interesting is the role that interactive video can play in all manner of training methods and the place it takes among the other major instructional media. As we will see in this chapter, interactive video is a tool for self-instruction, for tutoring and (perhaps surprisingly) for the classroom. Moreover, it shares similarities with many different kinds of media. It is video, but it can also be text. Some people like to think of it as a 52,000–frame slide projector, but it also shares many of the characteristics of programmed instruction text and computer-based training (CBT).

Interactive video can go beyond all of these media to teach subject matter which up until now has only been available in the classroom. Foremost among these subject areas is interpersonal relations.

In this chapter we will look at the various uses of interactive video, the subject matter it can teach, the situations in which it works best, its relationship with the computer and CBT, and some of the newer design strategies that have been applied to it.

First, let's review some of the subject matter areas to which interactive video training can be applied.

APPROPRIATE SUBJECT MATTER

Discrimination, generalization and sequencing are commonly known subject areas in which both programmed instruction and interactive video training work well.

Discrimination Skills

To discriminate, means to recognize differences—especially between similar objects. Discrimination skills, therefore, are essential for analysis and diagnosis. Mechanical troubleshooting and medical diagnosis are perfect examples of how these skills can be applied.

Video instruction provides everything that printed programmed instruction does with the added element of motion (and sometimes sound). This enables the user to not only identify objects by their appearance, but also by the ways in which they move. (This ability can be especially important in medical training in which students observe the internal functions of living organisms.)

Effective interactive video training for the development of discrimination skills employs a shaping process using images of increasing realism. Realistic images are harder to discriminate than, say, artists' conceptions that are drawn to exaggerate differences. In such training, students are asked to differentiate between actual items and those that only *look* similar. By moving from very exaggerated artistic illustrations to actual photographs of items, students can sharpen their skills at recognizing specific objects or conditions. For example, medical students can learn to differentiate between different kinds of tumors, thereby sharpening their skills at diagnosing certain medical problems.

Generalization Skills

Generalization skills are the reverse of discrimination skills. To generalize is to find the common rules or conditions that apply to seemingly diverse items or concepts. These include the identification of patterns, principles and general rules. Generalization is at the heart of conceptual learning and, like discrimination, is a valuable tool in many areas.

Guided discovery and inductive learning are generalization techniques particularly suited to interactive video. The ability of the user (or trainee) to move around within a video program, to deduce a set of rules or conditions and to arrive at a determined outcome cannot be matched by printed programmed instruction.

The following is a list of areas in which generalization training can be useful:

- Marketing—a sales strategy is determined by user demographics.
- Medicine—a series of symptoms add up to the identification of a particular disease.
- Criminology—a crime is solved by the gathering of clues.
- Philosophy—certain natural facts lead to a particular universal truth.

Sequencing Skills

The development of sequencing skills is the essence of procedural training. Often, it isn't enough to just apply a certain set of rules, they have to be applied in a particular order. Many "how-to" lessons have a sequencing component to them. The following list provides a few sequencing commands that can be used in a video training program:

- Where do you start?
- Which step is out of order?
- Put these steps in the correct order.
- Given this problem, what is the most logical order for troubleshooting?
- *You* tell us the order of events—*we* will show you what will happen.

Implicit in the last example—You tell us the order, we'll show you what will happen—is the essence of successful training with interactive video. What takes place here is simulation—simulation of the consequences of decisions, simulation of the thought processes that must be developed to obtain a particular outcome and simulation of the physical processes that must be exercised to make something happen.

Developing Psychomotor Skills

By simulating certain activities, psychomotor skills can be developed with interactive video. In other words, if you want to teach someone how to type, simulate the typing experience in a highly controlled situation by using the computer keyboard as the typewriter keyboard and the videodisc as a sensitive tutor. If you want to teach hand-eye coordination, simulate the desired behavior by using video images and a specially designed controller that looks and feels like the object to be operated. If you want to teach someone how to find his/her way through the streets of an unknown city, though it is not exactly a psychomotor skill, take the person for a simulated walk up and down the streets of that city using the cursor keys on the computer keyboard to control direction.

TRAINING PROGRAMS THAT ARE IDEAL FOR INTERACTIVE VIDEO

Interpersonal Relations

If professional instructional designers were asked to pick the subject least likely to be taught effectively by programmed instruction, they would probably choose management and interpersonal relations. After all, perfect rules for getting along with people, let alone managing them, are almost impossible to spell out in the detail needed to create good programmed instruction. Nevertheless, interpersonal relations training is one of the most popular subjects committed to interactive video. You might wonder why.

One reason is that while you cannot teach the *perfect* interpersonal relations system, you can teach *a* system. And, teaching *some* kind of interpersonal relations system is better than teaching no system at all. If you can specify a plan or approach that would give students a fairly reliable means of dealing with people, then you have gone a long way toward making them more effective managers, teachers or salespeople, whatever the case may be. So, accept the fact that the ultimate and definitive interpersonal relations system is beyond our grasp and settle for teaching a workable (though somewhat flawed) approach.

Interactive video can combine the powerful instructional logic of programmed instruction with the drama of real-life situations to create consequence remediation (in which you can see the results of your decisions, for good or for ill).

The dramatic elements that interactive video adds to programmed instruction bring it to life, increase student retention and transference back to the job. Interactive video exercises on interpersonal relations come closer than any other instructional tool to simulating a difficult interpersonal communication without causing the learner to actually suffer the real-life consequences of a mistake. It's possible that some day robots might act as better simulation devices. In the meantime, however, interactive video provides us with a training system that (at least in our opinion) is far more effective than simple role playing, and far less dangerous than sensitivity training.

Simulating the Use of a Computer and Its Software

Another application of interactive video is the simulation of the use of a computer and its software. Interactive video equipment has the device it would be simulating (the computer) built right into it. Yet, this application has not been exploited very much at all.

Many designers of interactive video programs argue that computers can simulate the use of their own software without video, so why simulate something that can generate its own simulations? The answer is obvious to anyone who works for a company that manufactures computers—the computer is not yet accepted by all users since a large part of the population does not relate well to computers. In fact, a target market for many computer manufacturers is comprised of people who *refuse* to relate to computers. The bottom line is that most people would rather learn from a video screen than from a computer screen. So, using interactive video to teach computer usage makes good sense.

As we get into the specifics of interactive video exercises we'll see exactly how interactive video can go beyond basic computer based training (CBT) to provide a more complete learning experience in this area.

FORMATS FOR INSTRUCTIONAL EXERCISES

There was a time, only about 15 years ago, when many instructional designers believed that two or three exercise formats (multiple-choice and fill-in-the blanks) were enough to teach anything. This misconception was carried over into the design of interactive video exercises.

Fortunately, the trend has changed and interactive video designers are now coming up with more and more interactive video exercises. This is the way it should be since the cardinal rule of exercise design is to focus on the behaviors to be learned and simulate them in the best ways possible. With a wide variety of behaviors being taught it only stands to reason that a wide variety of exercise styles should be used to simulate them. To put it another way, there are only so many things that you can simulate with multiple-choice questions.

The biggest problem with the original misconception—that you can teach nearly everything with one or two exercises—is that it was believed by the designers of interactive video authoring languages, who consequently turned out those menu-driven packages for generating a control program for the video and all the computer text slides to go with it. The original authoring languages offered only one or two exercise format options to the users who bought them. A typical program design choice is illustrated in the fictional menu frame in Figure 3.1.

**Figure 3.1: Program Design Menu Frame From
Archaic Authoring Language**

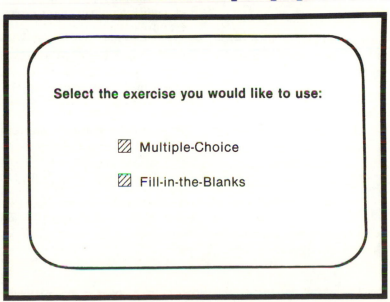

There has to be more to interactive video design than a choice between these simple exercise formats. We will talk more about authoring languages and the options they offer to instructional designers in Chapter 11. For now, let's make sure that the people trying to sell us authoring languages know that we want the largest number of exercise formats possible in our programs. Tables 3.1 and 3.2 list many different types of interactive video exercises. Try some of these when the need arises.

Table 3.1: A List of Interactive Learning Exercises

Type of Exercise	Exercise Name	Procedure
Interruptive	Answer and Compare	Students are brought by the video to a point at which they must perform an activity. The video poses a question, then stops and waits. Students answer the question in a workbook, on a sheet of paper or on a computer. When students have completed their answers, the video presents the correct answer. Students compare their answers to the answer on the screen. Extensive feedback may also be provided with the correct answer.
	Guided Discovery	The video image presents instructional material about a particular subject. Students are asked to repeat what they have just seen. For example, a video providing instruction on the use of computer software would indicate the key to be struck to accomplish a particular command. The video would then wait for students to hit the appropriate key on their terminals before proceeding to the next instruction.
Identification	Stop When You See	The video displays a series of activities. Viewers are then given a scenario for which one or several of the activities are appropriate. When the correct activity is recognized viewers strike a key.
		The "spot the shoplifter" scene in David Hon's *Demodisc* may be one of the best examples of the stop-when-you-see technique. David presents a store with four customers. One of the customers is a shoplifter. Using techniques that have been previously taught, students are asked to identify the shoplifter when they get the appropriate clue.
		There are several ways to get feedback in an exercise such as this: from audio track 2, from compressed audio, or from feedback sequences placed at the end of the entire program segment.

Table 3.1: A List of Interactive Learning Exercises (Cont.)

Type of Exercise	Exercise Name	Procedure
Identification (Cont.)	Identify the Area	This great tool for teaching medical diagnosis techniques requires computer graphics overlay. Students have an arrow (or other marker) they can control with a mouse, trackball, joystick or the cursor keys on the computer keyboard. Students move the arrow over a background until they see something they have been asked to identify—a tumor, a fracture, etc. Students then hit another key on the controller, when the arrow is pointing where they want it, to get appropriate feedback (again, from the audio channel, from video images or from the computer overlay).
		This is a great technique to use with touch screen systems.
	Matching	The best application of matching we've seen is Wilson Learning's *Managing Interpersonal Relations* (MIR) interactive program. A computer text overlay is employed that presents a list of personality types. The video shows pictures of four people students have already met in the program. Students first touch one of the listed personality types with a light pen, then the picture of the person who matches that type.
		Of the available feedback techniques, the one The Wilson Group chose may be the best—the person identified appears on the screen and tells students why they were right or wrong.
Sequencing	Put in Order	Students are presented with several steps that are listed as text or, better yet, portrayed as insert images on a composite picture. Students are required to indicate the order in which the steps must be performed—they can touch them in the right order (on a touch screen) or enter the numbers or letters identifying the scenes into the computer. Feedback can illustrate the resulting consequences of doing things either in order or out of order.
		This type of exercise is a must for procedural training that requires sequencing skills.
Multiple-Choice	Remedial Loops	Remedial loops are the granddaddy of all interactive video exercises. In this technique, students are asked a question that has four (or fewer) possible answers. If a student selects the correct answer, the exercise moves along. If the student selects the incorrect answer, the program loops back to the part of the lesson relating to that particular question.

Table 3.1: A List of Interactive Learning Exercises (Cont.)

Type of Exercise	Exercise Name	Procedure
Multiple-Choice (Cont.)	Remedial Loops (Cont.)	While it is tempting to use this technique because it is inexpensive and no additional production is required, it is weak because the feedback is not tailored to specific questions or responses. (In our experience, general feedback is usually not as effective as specific feedback.)
	Individualized Feedback	This technique is similar to remedial loops but much more effective. Students are asked questions with four possible choices, as with remedial loops. In this case, however, when students answer, they move to a block of video that explains *why* they were right or wrong.
	Consequence Remediation	Consequence remediation is similar to individualized feedback, however with this technique, when students answer they are shown the results of their decisions. There may be a voiceover explaining what happened, but the real power of this exercise comes from students seeing the consequences of their actions.
		Example: You are a sales clerk in a department store. An angry customer comes up to you and begins a conversation. You decide to respond with anger of your own. What will the customer do? The video shows you—watch out!
	Randomized Outcomes	Randomized outcomes provide yet another twist to the multiple-choice format. In this exercise, a randomizer is put into the program so that the consequences of certain actions are not always the same.
		For example, in an interactive video game in which students answer multiple-choice questions in competition with an on-screen opponent, students go to correct feedback when they answer correctly. When they answer incorrectly, the randomizer takes over and decides whether their on-screen opponent gets the answer right or wrong. (For more on randomized outcomes, see Chapter 5.)

MORE INNOVATIVE APPLICATIONS

Table 3.1 lists interactive video applications that are relatively conventional. Now, let's consider activities that are on the leading edge of interactive techniques. Table 3.2 lists interactive video exercises that expand on the instructional formats we have discussed to provide improved simulation.

Table 3.2: More Innovative Applications

Type of Exercise	Exercise Name	Procedure
Database Usage	Check Your Files	Database usage is an innovative technique that has been used effectively at Hewlett-Packard. This technique involves the creation of a large database of information that is stored at a central location on a videodisc. The information is organized in a logical order—however, the order should not be too easily discernable for students.
		Management trainees, for example, may be presented with a problem employee who needs counseling. A low-level simulation would provide students with a menu that offers to show the employee's personnel file and any recent mail that specifically relates to the employee in question.
		A high-level simulation would provide files for many different employees. There would also be mail for the last few days. The intention here is to get students to dig harder for the information they need.
		Is this second version of the exercise more difficult to create? Yes! Is it more effective? Evaluation conducted by Hewlett-Packard suggests that it is.
		A management course, also designed at Hewlett-Packard, taught that long-term relationships between employer and employee are factors in choice of management style. To demonstrate the length and depth of the relationship, Hewlett-Packard's course offered a memory bank through which the manager could reflect on the relationship before the counseling session actually began. A simple series of flashback scenes did the job. Students were able to determine the appropriate style with the help of this input.
	Market Research	Here is an especially effective technique for use with students in marketing courses. Offer trainees files of demographic data that they can draw on when putting together marketing plans. Sources for this database might include statistics from company marketing research departments, recent articles from major magazines and interviews with industry experts, which the students can draw upon as input.
		The easy and obvious exercise for this kind of database would be to supply a choice of several different advertising campaigns for a particular project, with resultant remediation showing the consequences of each choice. More imaginative or

Table 3.2: More Innovative Applications (Cont.)

Type of Exercise	Exercise Name	Procedure
Database Usage (Cont.)	Market Research (Cont.)	complex solutions could be carried out in the classroom. There, students could be organized into teams to design advertising campaigns. They could see the consequences of their campaigns in scenarios selected as appropriate by the course instructor.
		For a more detailed discussion of this example, see Chapter 4.
Pathfinding	Parallel Scenarios	Parallel scenarios offer students a variety of decision points leading to different solutions to the same problem. At each of the decision points, students are given another set of four differing choices. Each choice allows students to switch between paths and approaches. Each path uses similar data but carries it out in a very different way to arrive at a unique outcome.
		Example: If an employee acts with anger, how should a manager respond? Should the manager respond with anger also, or try to calm the employee down by telling a joke or empathizing with the employee's situation? Any reaction on the part of the manager results in another reaction on the part of the employee. For each response, the manager is faced with yet another set of choices, each leading to a very different set of outcomes.
		From the first decision point on, there are four parallel conversations going, each involving some of the same data, but each with its own special tone and style depending on the tactic chosen. Crossover points—points at which another counseling style can be accessed to continue the conversation using a different tack—are placed at logical breaks in the conversation where it would be natural to change direction. These points are dictated by the content and context of the conversation.
		A parallel program must be created with very carefully written dialogue so conversations can flow naturally from one scene to another. The emotional tone should also flow, even though there might be abrupt changes in mood. In order to accomplish this, transitional scenes may have to be created so that a highly charged emotional scene in one scenario can flow into a more amiable scene that results when another tack has been elected.

Table 3.2: More Innovative Applications (Cont.)

Type of Exercise	Exercise Name	Procedure
Pathfinding (Cont.)	Seamless Parallel Scenarios	Parallel scenarios that are not interrupted by menus or other decision-making devices are expensive and difficult to design and produce. Yet, many interactive game companies are exploring these techniques in their research and development units. The extremely high level of simulation offered by such scenarios makes them good prospects for training as well as for consumer games.
		Since there are no menus, the interruptions cannot be seen and the transitions are said to be seamless—no one even knows they are there.
		When you consider the difficulty of designing and creating a scenario in which the decision maker can switch the tone and mode of the conversation at a dozen or so different points, you can imagine how much more difficult it would be to create a scenario in which the decision maker can switch at any time—at will!
		The way to write such a scenario is to lay the scripts down side-by-side and write each scene so that it is in total harmony with the spirit and the informational content of the other scenes. This is not an easy task, and the writer who can do it should rightfully be considered quite an artist. But, that's another issue.
		Currently, the only way to create a videodisc that uses this high level of simulation is to use the technique of interleaving.
	Movie Mapping	Similar to MIT's pioneer project that offered a tour of Aspen, CO, movie mapping projects enable disc designers to offer viewers a series of scenes that take them up and down the streets of a town (real or imagined). Or, with very little change to the design, up and down the halls of a building (real or imagined).
		The technique allows the user to decide which way to turn at appropriate intersections and even whether or not to go into buildings or rooms that are found along the way.
		By using a computer image insert (or a second screen) the system can track a user's path through the environment.

Table 3.2: More Innovative Applications (Cont.)

Type of Exercise	Exercise Name	Procedure
Pathfinding (Cont.)	Movie Mapping (Cont.)	Movie maps offer training in recognizing landmarks as well as other kinds of geographic skills. They offer orientations to specific locations, buildings and building complexes, and can provide the raw material for a number of exciting adventure games.
		As a demonstration, Hewlett-Packard created a guide to the labyrinth of its offices by putting a 35mm camera on a chair and rolling it up and down the corridors. The hundreds of still frames that were created could be arranged in many ways, including a "You type in the name and we'll show you how to get to the office," routine.
		The potential of movie mapping for orientation, gaming and geographic/map reading training is great indeed.
Psychomotor Simulations	Hunt and Shoot	The most basic video games are those that require players to seek out a target and shoot it down. Videodisc versions of such games use the videodisc to create a highly detailed background upon which computer graphic images are overlayed. Many of the early videodisc games used computer-generated aircraft as targets. (In one particular game, alien spacecraft were assailed.)
		Psychomotor simulation was encouraged by the reflex action needed to steer a player's aircraft (usually with a joystick), and the hand-eye coordination required to shoot down enemy aircraft, which usually appeared with great suddenness.
		Some advanced interactive video game designs featured videodisc-generated monsters that could be shot down as they popped in and out of the landscape. These monsters could usually fight back and games ended when the monster destroyed the game player. Often, to give the user something to do while the monster was out of sight, computer-generated beasties of a much smaller variety would slither across the outer edges of the scene to add to the shooting gallery appearance of the game.
		Such ultra-sophisticated videodisc games might have maintained the booming popularity of video games, but the game manufacturers and designers killed the goose that laid the golden egg by turning out much less sophisticated videodisc games that were just not fun to play.

Table 3.2: More Innovative Applications (Cont.)

Type of Exercise	Exercise Name	Procedure
Psychomotor Simulations (Cont.)	Driving Simulations	This hybrid of the video game and the movie map never really got off the ground. It was perhaps one of the very best applications of interactive video-disc simulation. The player would take the controls of a futuristic car and the disc images (à la movie map) would allow him or her to traverse the streets of a major city—San Francisco was ideal.
		A storyline could be overlaid on the basic activity so that adversary cars could chase the player. Hunt and shoot controls added to the excitement.
		(The only way to produce truly effective inter-active video driving games is to use the process of interleaving.)
	Computer Software Simulations	Some computer software applications require a degree of psychomotor skill and, to that end, we need merely add that most computer software programs have built-in hooks to allow for the creation of computer simulation programs. The trick is to get at those hooks so they can be used to control a videodisc player that demonstrates what to do, what happens when you do it, and then presents practice exercises with increasing degrees of difficulty.

INTERACTIVE VIDEO AND COMPUTER-BASED TRAINING

The discussion of interactive software simulation brings us to the issue of computer-based training (CBT) without video. For a long time, computer text was used to effect the types of activities we have been describing in Tables 3.1 and 3.2. With the growing quality of computer graphics, it would appear that computers can do as much as video can. This leads us to the question: Are CBT and interactive video competitors, or can they be used together?

The answers seem obvious. Interactive video should be an adjunct, rather than a competitor to CBT. But often, when training programs are being put together, in the heat of battle, there is quite a struggle to see which component will dominate the learning experience.

We have witnessed situations in which training has been computer-based with a little video on the side. We have also seen examples of interactive video in which the computer has been relegated to functions similar to those performed by a numerical key pad.

There has to be something better than these two extremes. There has to be a better way to marry these two technologies.

The key to successful integration of interactive video and CBT is to allow each element of the integration to do what it does best. The assignment of responsibilities varies somewhat depending on the skill being taught and will be especially apparent in the interactive learning exercises. However, some basic divisions of labor are apparent due to the very essence of the interactive video process.

The Role of Video

Presentation of Information

Presentation of information should almost always be done on video. Video brings sound, color, motion and many other elements to the task of presenting information. This is important because computer-literate developers are often tempted to start by writing several blocks of copy to welcome students to a program and tell them how to use it. We believe, however, that all the welcomes, how-tos and instructions should come from the video. It will be easier for the viewer.

We are not trying to discourage literacy here. Our job is to get people to learn, to use the instruction and to assimilate the information. Reading long paragraphs of instruction can be a barrier to these objectives.

Feedback

Remediation feedback should also come from the video segment whenever possible. Even if an exercise is carried out with a computer keyboard, the reinforcement for the right answers and the remediation for the wrong answers gain power and effectiveness from the video image. Reinforcement, after all, is a reward for getting the correct answer. Being *told* that you got it right can be a key ingredient in that reinforcement.

The video could be reinforcing

When we look at feedback from the point of view of the remedial segments, the need for video feedback becomes even stronger. Remediation is the corrective feedback—"This is why your answer is wrong."

Video feedback can illustrate why an answer is wrong and can explain the principles that should be at work during the decision-making process. This can often require lengthy communication; and clearly, lengthy communication is the forte of video, not computer text.

The Role of the Computer

What, then, does the computer do? The most obvious answer is, control.

The computer tells the videodisc player what to do. Its second and less obvious use is to act as a response mechanism. Students answer questions by using the computer keyboard or other control mechanisms. More creatively, the computer provides the means by which students can simulate behaviors required by different exercises.

Simulation Enhanced by Computer Graphics

In a program on how to hardwire a modem eliminator cable, students learn how to connect wires from pins on one end of the cable to pins on the other end. The course is not about soldering, so the mechanical act of making the connections is not important. Students simply have to learn where the connections should be made.

Completion of a wiring diagram is a perfect simulation exercise. The exercise can be done in a workbook, but when it is practiced on a computer, the immediate video feedback and color graphics capability of the computer create a much more effective learning activity.

Of course, the activity in this example and many others is enhanced when the advanced graphics capability of the computer is mixed with video. In our example, a video still of the pins and wiring points of two ends of a cable could be joined with different color lines selected by the student and created by the computer graphics system. The lines would indicate which pins should be wired together.

how done ?

In addition to laying computer images over video backgrounds, advanced computer graphics provide the ability to make what appear to be computer-generated backgrounds. During this process, the computer generates a page of color on top of the video images. The small video images show through holes or "windows" in the computer-generated page to create the illusion of images set on top of a background color.

One of the most impressive examples of this windowing technique is the IBM *Info Window System*, which offers quality computer graphics mixed with videodisc images that are controlled by a touch screen.

Conventional Testing

Conventional testing can be completed at the end of lessons or units. These tests use text only and are, therefore, usually computer-generated. Though conventional tests are not necessarily the ultimate testing tool, they offer a solid simulation of the testing experience and can be a great benefit to students, especially in courses where written tests are required for certification.

Teaching the Use of Computer Software

Teaching computer software use is a subject area in which the computer plays an expanded role in the interactive video process. What happens on the computer screen is the subject matter of these courses, so students should see every possible example of how a program can be used during the computer-based training (CBT) lessons. Even though many of the exercises will display samples of computer screens, it is still important to convey the bulk of the instructional message through the video and audio components of the videodisc. The video's ability to zero in on one small part of the screen, to highlight areas, to provide split screens for comparison of several images and to give audio as well as video input can greatly improve the quality of the learning experience.

For those who feel the computer is unfriendly, video images of real people who act as guides throughout the lesson can be a welcome change. This humanization of software training could help to expand the use of CBT/interactive video systems to people who would prefer not to be computer literate.

TAILORED APPLICATIONS OF CBT/VIDEO SYSTEMS

A popular instructional theory suggests that it is not always necessary to teach people everything there is to know about a given procedure or skill in order for them to use it effectively in their jobs. In fact, it may be more effective to teach people only parts of a procedure—those parts they will use most frequently.

If, for example, you are designing a course that teaches people how to use a piece of software that files and cross-references information, and the people you are going to teach are financial counselors, it might be best to focus only on those parts of the software that would be used most by financial counselors. In this way you can devote your training time to repeated practice of these specific parts of the program (such as creating files of customers and listing their financial demands, creating files of financial products and their characteristics, and then matching customer needs to product characteristics).

This format of program development will allow you to skip the complications and confusion that come from teaching the irrelevant information that is often included in general theory programs. A general theory of filing, for example, may teach a filing system that in no way supplies the kind of information needed by financial counselors.

As you may imagine, this approach is not universally accepted by instructional designers. There are still those who feel that a broad orientation to a subject (sort of a liberal arts approach to training) is the very best way to begin any learning experience. But more and more people are beginning to believe that tailoring instruction to the activities performed by a carefully defined target population improves training effectiveness and speeds transfer of skills from the classroom to the workplace.

One of the nicest features of interactive video from a designer's point of view is that it provides the ability for compromise on issues such as tailored training applications. Interactive video allows for the inclusion of material on the videodisc that may or may not be used. This means that you can tailor some applications to specific learners, while also building large parts of the program to provide more general information. The net effect is that the learner can decide which learning technique he or she prefers. If one woman learns better with a general overview, offer her that choice. If she learns better by getting practical "hands-on" instruction before beginning the general exercises, that's a possibility too. If she does not know or care which type of instruction she receives, devise a test that makes the decision for her.

The ability to tailor applications to a target population makes it possible to provide focused, accurate and more cost-effective training. The unique characteristics of the interactive system make it the ideal medium for such instruction.

INDIVIDUALIZED INSTRUCTION VERSUS GROUP INSTRUCTION

At the Bank of America in 1980, a very interesting discovery was made. The bank was developing a new line of interactive video teller training programs. The concern was that the new technological effort would not be supported by the teller–school program and its instructors.

The bank's solution to the problem was to develop the first module on a subject that was so complex and unpleasant to teach, that instructors would be grateful to have some media to support their instruction of it. When the module was finished, it was given to the teachers for use in the teller–school classroom setting. An amazing thing happened—not only did the teachers become some of the strongest supporters of interactive video technology, but the system turned out to work very well when used in a classroom environment.

You may question whether or not the entire purpose of individualized instruction is defeated when an interactive videodisc is presented to a class of students who go through it together in a lock-step fashion under the guidance of an instructor. The answer is both yes and no.

Even though students do go through the exercises together, those students who do not completely grasp the material during the group activity can go back at a later date to use the program individually and at their own pace. This is exactly what the Bank of America students did. Those students who did not understand the procedures presented on the disc showed that deficiency when they entered the "hands-on" part of the training. These students were sent back to the interactive video program to go through it at their own pace.

Classroom instruction is an established and effective way to teach. Interactive video goes very far in improving that method. In its simplest form, interactive video can be described as a 54,000-slot carousel projector that allows an instructor to show video images of key points, examples and samples, many of which can be moving (if that is required). At its very best, interactive video and the teacher are partners in a dynamic and effective presentation—much more dynamic and effective than, say, filmstrips or even linear video in which the teacher shows a whole program and the students watch passively. Video can give structure, order, excitement and even entertainment value to a classroom presentation. The teacher's job is to make up for the lock-step nature of the situation by getting students involved. The teacher comments, interprets, enhances and, oh yes, *controls* the interactive video presentation.

Video provides organization and a clear-cut presentation of information for students. Nothing can be left out of a program or be interpreted incorrectly—which is something that all designers of curricula worry about. At the same time, teachers can make up for deficiencies in the instructional design of a training disc by taking poor examples of principles or practices and making points about them. In a recent management training disc, we noticed that an actor's body language while portraying a manager counseling an employee was less than ideal. On the other hand, it was very much what you would expect an untrained person to do under such circumstances. An instructor might choose to elaborate on this difference by indicating that would-be managers should be especially careful about their body language during counseling sessions.

Since most interactive video material has been designed for individualized instruction (because of the tremendous cost savings), we have yet to see the full advantages of interactive discs that are designed primarily for use in the classroom. Several designers, however, ourselves included, are currently involved in designing programs for just that use. We feel that the benefits will be great and that discs that work both as classroom tools and as individualized instruction systems will soon be recognized for the tremendous improvement they can bring to the learning environment.

CONCLUSION

Interactive video is a powerful teaching tool, one that may turn out to become part of the greatest learning system yet devised. If interactive video is to reach its full potential as a training medium, however, it is important that program designers keep their training objectives paramount in their minds. They cannot become so enamored of interactive hardware that they forget what they are about. If the exercises and the technology are made to match, then they cannot fail.

4 Interactive Video in Education

Interactive video has become a valued educational tool. While it would be nice to present an extensive list of interactive programs that have been developed for all levels of education, most of the work that has been done has centered around the elementary school and university levels.

At the elementary level, companies such as Apple Computer, Inc. have been working in conjunction with major film and television producers such as Walt Disney Productions, Lucas film and the National Geographic Society to build prototypes of advanced instructional media.

In universities, schools of engineering and business have often worked in conjunction with major corporations to develop interactive teaching modules, as well as experimental media applications using interactive video. (MIT's Media Lab is most notable in the latter category.*)

This chapter will focus on interactive video as a tool to be used in education. Stress has been placed on its use in elementary schools where so much of the work has been done.

ELEMENTARY EDUCATION

Consider the carefully crafted, well researched and brilliantly successful PBS television series *Sesame Street* and the approach it takes to teaching children. The format of this series presents information to children in a manner they like. It is bright, loud and fast-paced with interesting graphics, lively music and entertaining characters.

---

*Note: See Stewart Brand, *The Media Lab, Inventing the Future at MIT* (Viking Press, 1987).

Studies indicate that because most children watch a great deal of network television they have learned to accept information in short bursts— about the length of a TV commercial (30 to 60 seconds). *Sesame Street's* strategy for teaching children has been to imitate this process by presenting information in small segments. The more a message looks and sounds like a commercial, the more tuned-in children seem to be.

Sesame Street also employs repetition and mediation (songs, rhymes and jokes) to help children remember. When it comes to instructional strategies, the brilliance of *Sesame Street* may be that it uses a smorgasbord approach to education (lots of everything). That may be the best way to go for both educational television and educational interactive video.

THE ROLE OF THE TEACHER

So, what's missing in the televised learning experience? What can you find in a classroom that you cannot find on *Sesame Street*? The answer is obvious—interaction with a real human being—a teacher.

Excellent teachers are still the best educators. Few educational technologists will argue with that. Most of us have probably had a teacher whose classroom lectures and presentations were so vivid that they surpassed those of the best video presentation. But, not all teachers have great presentation skills.

The use of instructional aids in the classroom, however, can help to make every teacher's performance more effective. For example, teachers who lecture about complex subjects can use media to dramatize certain aspects of their discussions and teachers who work with slow learners can use interactive video to help them with the remedial and tutorial aspects of their work.

Of course, teachers will always need to handle all the interpersonal activities— the people business—the hands-on demonstrations, private tutoring and classroom management, as well as the motivating, the caring and the disciplining.

The great high-tech teachers of the future will have more tools at their disposal to bring subjects alive for students. In the following discussion we will look at what a high-tech educational system might be like, a system in which high technology and teachers work together in the classroom.

THE CONFIGURATION OF THE INTERACTIVE VIDEO CLASSROOM

There are two ways that interactive video systems can be used in the classroom. They can function as a total learning system, or as components in a larger classroom experience.

The Total Learning System

In the total learning system the computer deals with each student individually and provides all of the demonstrations, exercises and testing. The classroom instructor's role in this process is as administrator, special advisor and, most important, as tutor. He or she oversees the assignments, provides a framework in which the students work, and deals with individual students as they make their way through the material in the learning system. When group activities are called for, the instructor leads the group activities.

The configuration for such a system is almost always separate learning carrels for each student. The carrels consist of a monitor, personal computer and a disc player. The computer presents the information in small *Sesame Street*—like "bursts." The computer provides for student practice and testing.

In the most elaborate example of the hardware configuration for a total learning system, students sit at separate learning carrels to interface with terminals that are tied into a central computer. The central computer holds the operating program that allows for record-keeping (test scores, student progress checks, etc.). At the students' commands, the central computer downloads the control program to a PC that is used to control the videodisc. (PCs are used because they are well-suited to the precise individualized control that is needed to run a finely-tuned interactive videodisc program.)

Each student watches his or her own screen and responds by touch (touch-screen) or with a keyboard or mouse, depending on the requirements of the program. (See Figure 4.1.)

The software for total learning systems is currently quite limited but is becoming more and more readily available. Programs are often prepared by large media producers. Many of the programs under development offer the option of teacher modification so that teachers can make the program work with their *styles* of teaching and the *requirements* of their curricula. With "teacher authoring" built into the design of the material, teachers actually have more control of the classroom situation than they have ever had with filmstrips, motion pictures or linear videotapes. These programs can be used in a total learning system or modified to become a component in a larger classroom experience.

Interactive Video as Learning Component

One of the current educational buzz-words is multimedia. The term has been around for some time, but, more and more, it is becoming synonymous with interactive video since interactive video is, by its very nature, multimedia (it requires both video and computers).

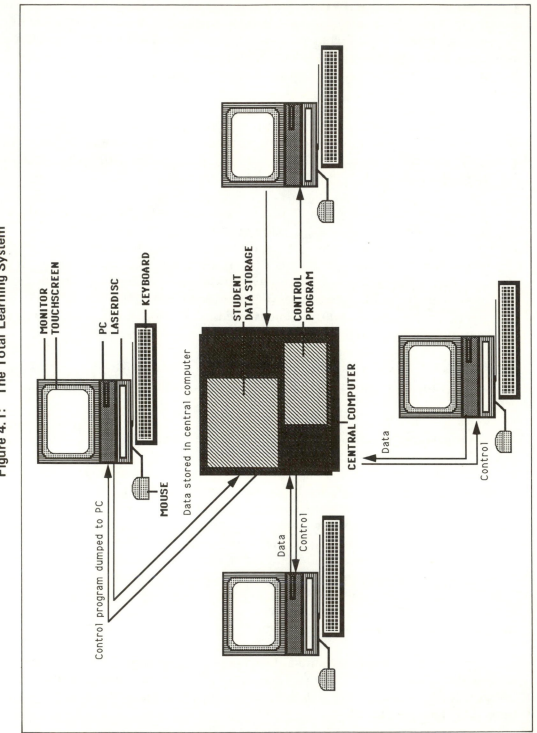

Figure 4.1:   The Total Learning System

Teachers who use interactive video as a learning component have many excellent videodisc programs already at their disposal, especially in the area of science. What has been lacking, but is soon to arrive, are a variety of authoring tools that will give teachers greater access and control of the disc program and will allow them to create customized audio/visual presentations for their classrooms.

Remember that the disc player holds 54,000 still frames or 30 minutes of full-motion audio and video (or some combination of the two such as 1000 stills and 28+ minutes of motion). The teacher, in essence, has a 54,000 frame slide projector, or a slide and video projector that allows media to be arranged in different ways for different lecture or demonstration purposes.

A teacher can go through a lesson plan, note the segments of a given video program that are relevant to a particular classroom discussion, prepare a menu that will allow for quick access to the segments and then have motion segments and still-frames to show the class during the next day's session. (Chapter 11 discusses the use of computers in menu preparation.)

Perhaps the best example of such a teaching system is that created by Hewlett-Packard in conjunction with the Harvard Business school.

The system consisted of three Sony 2000A disc players that were controlled by an HP Vectra computer. The subject was a marketing case study.

The video consisted of taped interviews with members of a corporate marketing team, related graphics, and articles from newspapers and magazines.

A PC was used to index the segments. This allowed instructors to use a touch-screen for browsing. When the instructors had selected the required segments they could be programmed to appear in a particular order for class presentation.

The system also provided instructors with the use of the full range of disc player controls (play, freeze, fast-forward, reverse), as well as memory of the segments. The system could even remember several segments in a row and create a video sequence by playing them together.

It is clear that this system is a *component* in the classroom experience. It has proven to be a powerful and effective tool however, and when used in the classroom, can add realism and depth to discussions, as well as give support to the instructor.

REPURPOSING EXISTING VIDEODISCS

One of the first things a teacher can do to develop his or her own interactive video materials for the classroom is to take existing video programs and create *computerized* control programs that are especially suited to the course and curriculum being taught.

Using, for example, a laserdisc player, a Macintosh with *Hypercard* and an existing CAV videodisc, a teacher can build a two-screen presentation tool. The computer screen shows menus, metaphors and parts of exercises. The video presents the demonstration material, special messages and the rest of the exercises. Here are some suggestions on how to use such a system:

Make a Metaphor

Create a menu that is a metaphor for a familiar way to get at information.

- Using *Hypercard* on the Macintosh, create a menu that looks like a sea full of islands (islands of information.) The Sea of Science might have islands for Physics, Astronomy and Chemistry (each programmed to correspond to segments on a science videodisc). By moving your cursor from island to island you can access various subjects on the disc.
- Create the front page of a newspaper which refers to various related historical events. If you have access to a scanner (a machine that puts pictures into a computer), scan in relevant photos. Add on headlines or parts of stories when you want to access certain information.
- Diagram the overall flow of information in a program. If the program is about the exploration of the planets, diagram the path of a rocket with pictures of the planets that you can visit as you go by.

Look for "Video Clip" Comments

As the writers and directors of commercials and coming attractions do, look for little clips which, when taken out of context, can serve as reinforcement feedback or special messages.

- The main character of a video says, "This is a very important mission." Identify the frame numbers around that statement and make it available whenever you want to say it to your class.
- A cartoon character says, "You've done a wonderful job." Present this statement to your students when they have done a wonderful job.

Put Reference Material into the Computer Section of the Program

Create a teacher's desktop that you can access with an icon button, located in the corner of the frame. The desktop corner can be filled with objects like pieces of note paper, notebooks or calendars. The symbols are really buttons that access teacher materials.

Build Creative Exercises Around Key Bits of Video

Maybe, your video program is a scientific presentation that explains different kinds of geological formations. Build exercises that ask your students to identify formations from still frames you select.

Maybe, there are other segments in the video that could be turned into discrimination generalization or sequencing exercises.

The secret of building good exercises is to study the video. Look at your objectives and see just how many interesting exercises can be created at the point where the video and objectives converge.

Exercise construction is, of course, the heart and soul of good interactive video. It is a challenge to the very best of teachers, but it is also one of the most creative things that either video writer, producer or classroom teacher can do.

INTERACTIVE VIDEO STRUCTURES

Now that we have looked at the two major interactive video hardware configurations and talked about ways that teachers can repurpose existing video programs, let us consider the different structures for interactive video exercises. These structures will probably be most useful to developers of total learning systems. Although, teachers who are willing to spend the time will be able to use them to design interactive components for their classroom activities as well.

Individual Modules

The most basic type of structure is no structure at all. Individual modules can be strung together so that they follow rapidly, one after the other without interruption or obvious order. Modules are generally stand-alone exercises of very short duration. This method of exercise structure is similar to the *Sesame Street* approach to teaching we discussed earlier.

Imagine a *Sesame Street* program in which every third or fourth segment of information is presented not as a cartoon or by a puppet, but as an interactive video exercise. For example, in a lesson teaching children to recognize the letter "Q," students would be asked to touch the letter "Q" when it appears on the screen. This can be done by either selecting the "Q" from among six other letters that are present on a single frame, or by catching it as it flashes by between different letters, numbers or other images.

The *Catch "Q"* exercise mentioned above can stand by itself. It can also be one of 15 other exercises about the letter "Q" that are available to be used whenever the teacher deems it appropriate. The modular approach to interactive exercises allows for the creation of a product that might be described as an *interactive video activity box,* which provides students with sets of exercises that can be chosen according to interest.

Demonstration Followed by Activity

This method of exercise structure begins with a short presentation of information on video. Exercises designed to reinforce student comprehension follow immediately. Often, the person or character leading the video demonstration also leads the exercises. The following instructions might be used by the demonstrator conducting an exercise on simple addition:

> "This is how to add $2 + 2$. . . . Okay, now *you* try. I will be
> here to help you if you get into trouble."

Integrated Lessons

Integrated lessons provide an entire lecture or discussion with various integrated activities inserted along the way. These lessons can be very effective if a famous person or character students recognize is incorporated into the presentation. For example, a hypothetical elementary reading lesson could be titled "Mother Goose Teaches Us To Read."

In such a lesson, Mother Goose could put together an hour-long class on reading in which she presents selected vocabulary words. As she goes along students work on individual exercises relating to these specific words. In another segment, the entire class could work together on similar exercises with additional words, and an integrated activity at the end of the lesson could bring it all together.

Storytelling

Storytelling is designed to help children understand the intention of a lesson by putting it into a context they can relate to. Storytelling also mediates information to make it easier to remember.

Often storytelling gets in the way of learning because there is too much story and not enough content, or there are distracting story points that obscure the content. Storytelling does, however, provide great help for dealing with certain types of subject matter. If the balance between story and class activity is kept even, there can be great opportunities for learning. It is important that there be a large number of interactions throughout the course of the presentation (as opposed to many interactions lumped together at the end of a long passive video sequence).

A great example of interactive storytelling, with exercises, is the *stop when you see a clue* structure that teaches recognition and discrimination skills in subjects such as science, reading and language.

In exercises developed from this structure, students watch a story unravel. Students have been told in an introduction to the exercise that when they spot certain kinds of objects (relating to the content of the course) they should stop the presentation and ask for an explanation. Usually, these programs have two objectives— students play at helping the characters in the story while gaining skills at the same time.

Games

The game show approach to learning exercises is similar to the approach of exercise modules. Learning games are not video presentations that are interrupted by or followed by interactions; games are *all* interaction.

Lightning Round is a hypothetical interactive video game that would be fashioned after the concept of flash cards. In the game of *Lightning Round,* questions and problems flash past the learner, who has to make a split-second decision about each.

Recognition lightning rounds might flash a series of multiple-image, multiple-choice frames by the learner in ten-second intervals. The time element adds to the intensity of the game and its fun. Future psychomotor simulations which might have names like Driving Simulator (driver's education), Zap the Bacteria (hygiene) or Practice the Concertina (sight-reading music) could build skills by approximating a critical part of the behavior that the student is trying to learn.

Figure 4.2 illustrates the flowchart and track layout for a complex interactive videodisc educational game titled *Around the Block.** The game is designed to teach map reading and directional sense. In the game, the player selects two gifts to give to friends on the block. He or she must then consult a map to find the best, and safest way to get to the friend's house.

The game uses five-track interleaving to provide seemless interaction. There are transitional sequences called "Walking Along." If the player has selected a gift for a

---

**Around the Block* is copyrighted © 1988 by Nicholas V. Iuppa.

FIGURE 4.2: Around the Block.

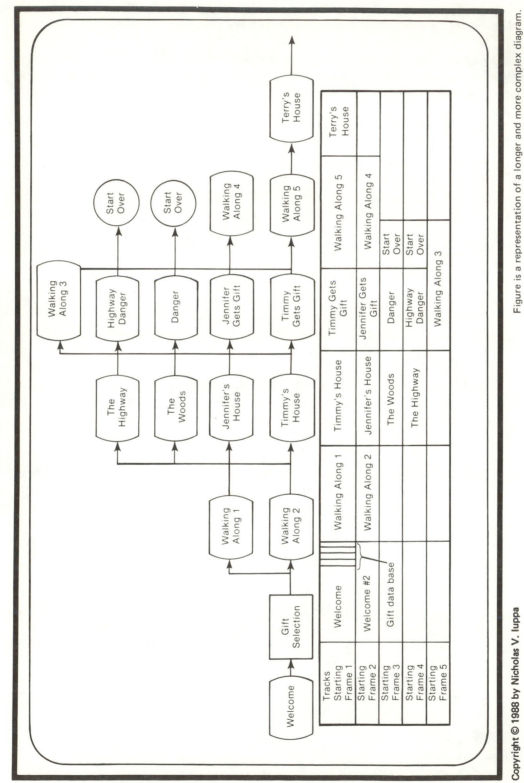

Figure is a representation of a longer and more complex diagram.

9.

person and gets to that person's house, the computer automatically plays a short and enjoyable gift giving sequence. If the player arrives at a house where a person does not get a gift, the player continues on along the sidewalk.

To add additional educational value, the block in the game is bounded by woods and a highway. If the player tries to cross the highway or go into the woods, he or she is warned of "Danger." If the player continues to cross the highway or go into the woods appropriate (but not too dire) consequences ensue and he or she must start the game over.

TYPICAL INTERACTIVE VIDEO LEARNING ACTIVITIES

Table 4.1 (see page 46) lists 13 different interactive video learning activities, as well as examples of each, the structural formats in which they work best and typical subjects that lend themselves to each of the activities.

CONCLUSION

The children going through today's educational systems are facing a flood of new and complex information that makes the job of learning as well as the job of teaching very difficult. It is fortunate that included in the flood of modern technology are tools for improved teaching. Today's teachers must take advantage of these tools.

This chapter has tried to suggest that interactive video, foremost among the new teaching tools, can fit into the educational environment in several ways. These range from total learning systems that take over most of the job of instruction, to learning components that can become powerful tools in traditional learning environments. In either case, the role and importance of the teacher is maintained and hopefully strengthened as well.

We have listed several types of learning exercises and the structures that they would fit into. Ideally, these concepts will stimulate the imaginations of instructional designers to develop software along these lines, as well as encourage instructors to seek out programs that bring innovation, fun and improved learning to the educational environment.

Table 4.1: Interactive Video Learning Activities

Activity Name	Example of Activity	Appropriate Structure	Subjects Suited to Activity
Matching	match picture to word	Modules	Reading
Counting	count symbols hit number key	Modules	Math
Fill in the Blank	put the vowel in the incomplete word	Modules	Spelling
Type it in	we say it, you type it	Modules	Spelling
Visual Multiple-Choice	pick the right one	All Structures	Science Spelling Math
Interrupt	stop when you see the clue	Games	Science Basic Recognition
Consequence Remediation	if you do this. . . this will happen	Storytelling	All Subjects
Drills	flash cards	Games	Reading Math Spelling Science
Lightning Round	flash cards	Games	Spelling Reading Math
Game Show	beat the expert	Games	Science Math Spelling
Spiderweb	discovery	Storytelling	Science Math
Optional Analogy	one way to understand is ____	Integrated Lesson	Math Science

5 Interactive Entertainment Styles

In the spring of 1986, an article in *Video Marketing* newsletter stated that:

> Interactive programming, which was once expected to carry the laser video format, has failed to live up to its promise.
>
> Sales rarely exceeded 10,000 units. Optical Programming Associates, once formed by Pioneer, MCA and North American Phillips to develop games and other interactive disc programs, was dissolved several years ago.
>
> ... According to Laserdisc Corporation of America's John Talbot .... "(they) won't introduce more (interactive) products for some time. The quality of the picture and sound is what's selling laservision. Interactive programming is complicated for consumers; it's the next phase."*

We do not agree with John Talbot's statement. Interactive programming is not complicated for consumers. Doing interactive video (which is really a Level 2 or 3 activity) on a Level 1 consumer player is impossible. Furthermore, attempts to develop interactive video for the Level 1 players have usually resulted in programs that are so clunky that they are difficult, unpleasant and even embarrassing to use. With a few exceptions (such as the First National *KIDISC* produced by Bruce Seth Green, © 1981 Optical Programming Associates) attempts to develop Level 1 consumer interactive video have merely confused the issue and made interactive video seem clumsy and ridiculous (like a tap dancer forced to perform in a suit of armor).

What is especially odd about these developments in consumer interactive video is that many of the efforts that have been made in the industrial and parts of the video game market do show what good interactive video can be. With little or no

---

*"Interactive Programming On Hold at LDCA, Simplified at MCA," *Video Marketing* newsletter, March 10, 1986.

modification, these designs can be put to work in the home entertainment arena to give consumers the benefit of interactivity. All this, of course, depends on the introduction of a home player that allows real interactive programming, and that's where John Talbot's words are correct. Interactive video home entertainment is the "next phase."

So, let us accept that the current success of consumer laser videodiscs rests on their use as a high definition medium for linear movie presentation. If we want to develop truly interactive consumer video programming, we are going to have to look beyond the material that is available on disc right now. We think that the material holding the greatest potential for consumer interactive video is the day-to-day programming of network television.

Many network programs are not only popular with the vast television audience, they already include elements that can be applied quite naturally to interactivity. Industrial video, ever-ready to imitate any popular concept used by network television, has already adopted many of these formats.

This chapter will focus on three extremely popular TV formats—game shows, adventure shows and soap operas. We feel that all three can be applied successfully to interactive video.

To show just how much potential they do have, we will take a close look at an example of each. Two of these examples have already been used as successful industrial interactive programs. The third is a proposal sitting on the desk of an interactive game producer right now.

We hope that these examples will stimulate interest in the development of interactive video for the consumer market as well as the hardware to make it work.

A SAMPLE INTERACTIVE VIDEO GAME SHOW

The Hewlett-Packard Company is currently using a product knowledge game show to demonstrate the capabilities of interactive video. The game teaches sales representatives about new products and helps customers learn about HP products and services. The basic design is one of the most fundamental and sound of all game show formulas, yet the program is so much fun that users frequently jump out of their chairs as they play the game. Many players talk back to the game show host and the opposing video contestants. One of the best advantages of the game is that it encourages customers to study, puzzle over and learn all the features and benefits of one of HP's best-selling computers.

Why is this method so successful? Because it is fun, of course—everyone loves game shows. The game show format is perhaps one of the best methods ever devised to test people's knowledge. It can even pass along a great deal of information in a pleasurable way. The flowchart of Hewlett-Packard's game show, *Know Your Product,* is illustrated in Figure 5.1.

How the Game Show Works

After the hoopla of the introduction of the host, the contestant and the rules of the game, the game show format boils down to a series of three-way multiple-choice questions, the calculation of the score, and prizes for the winner.

Getting the Players Involved

There is a trick to the mechanics of the game that makes a difference as to whether it will be a success or failure. The game is successful when the viewer (the off-screen contestant) gets immediately involved in the action of the game by getting the first crack at every answer. This is accomplished by giving the viewer ten seconds to answer before the question is offered to the on-screen contestant. If the viewer gives a correct answer the program branches to positive feedback from the host (see the flowchart in Figure 5.1).

If the question is not answered in ten seconds the viewer loses his or her turn. (The following "time's up" can be accompanied by a buzzer.) In such cases, or if the viewer gives the wrong answer, the question is passed to an on-screen contestant who is competing with the viewer for the grand prize. In industrial videos the prize is usually something small like a "product reminder card." Gold cards are given to the winners, silver cards to the losers.

So, the game now has three elements of interest for the contestant. The viewer is trying to:

1. get the right answer
2. beat the clock
3. gain more points than the on-screen contestant

Adding Complexity to the Game

There is another way, however, to catch and hold the viewer's attention—the on-screen contestant does not always have to be right. The on-screen contestant in *Know Your Product* was a woman named Margie. Margie was introduced as a "marketing expert." There were a variety of methods that could have been used to control Margie's ability to get the answer right.

The easiest way to direct the accuracy of Margie's responses was to let her be right *all the time*. Not very exciting but a good starting point since, as an expert, she

Figure 5.1: Flowchart for *Know Your Product* **by Hewlett-Packard Co.**

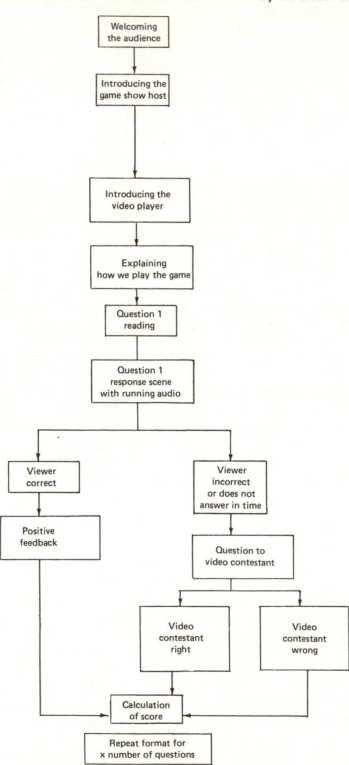

was *supposed* to be correct most of the time. The program designers, however, wanted to add a decent amount of incorrect responses from Margie to keep things interesting.

There were several ways to do this. For example, Margie could answer the majority of the questions correctly and the designers could pick the exact questions she would answer incorrectly. If it felt right for her to be correct 90% of the time the game's programmers could space out the remaining 10% incorrect answers at predetermined intervals to maximize viewer interest. (Of course, if the viewer played the game often enough, he or she would soon figure out when Margie was going to be wrong.)

Another way to determine if the on-screen contestant is correct would be to tie *Margie's* incorrect answers to the *viewer's* incorrect answers so that Margie would only be wrong when the viewer gave a specific wrong response. For example, let's say the correct answer to multiple-choice question No. 3 is C. If the viewer gave wrong answer A, Margie would give wrong answer B.

This is what the Hewlett-Packard designers did. It may not have been the most elegant solution to the problem, but it did work well for many reasons. First, it decreased the amount of exposure the viewer had to Margie's wrong answers. Margie would not always get the answer to question No. 3 wrong. She would only get the answer to question No. 3 wrong when the viewer had picked incorrect answer A.

If you think about it, this approach provides a better simulation of a real game between contestants than the other suggested solutions to the problem. Figure 5.2 provides a flowchart to illustrate the logic used to produce this program.

Figure 5.2: Basic Logic for Opponent's Responses

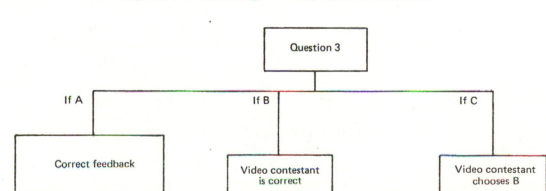

In the final version of the game show, Margie was correct almost all of the time (8 out of 12 times). In the four cases in which she was incorrect, she was incorrect only when the viewer had chosen a specific incorrect answer. The chance of Margie being wrong, therefore, was not 4 out of 12, but 4 out of 24. The viewer's chance of seeing Margie get the answer incorrect was 4 out of 36—just about the right ratio of correct answers to incorrect answers for an expert contestant.

Add a Randomizer

An alternative method of determining Margie's wrong responses involves the use of randomizers that flip the old "logical coin" every time Margie has a chance to answer a question. These allow Margie to respond with either a right or wrong answer based on the response made by the viewer.

For example, if the contestant chooses answer B to question No. 3, Margie would answer either correct answer A or incorrect answer C depending on the random selection of the randomizer. No matter how many times the contestant played the game there would be no way to determine whether or not Margie would answer correctly.

The program was able to use randomizers because HPTV had shot different "Margie" responses to every question in the game. These were put on the videodisc and allowed repeat players to play the game from a fresh and interesting perspective. Figure 5.3 shows the logic for a response mode governed by randomizers.

Consumer Game Shows

The popularity of game shows on broadcast television and the limited but very real success with interactive video game shows in industrial settings, suggests that a well-produced consumer game show might be suited to the entertainment market when (but *only* when) Level 2 or Level 3 interactive video is available for the consumer.

INTERACTIVE ADVENTURES

Much of the prototyping work that was done for interactive adventure shows took place in the research and development areas and the skunkworks of Silicon Valley companies like Atari and ByVideo.

Other prototyping work that applied to interactive adventure use showed up in medical training simulators, loss prevention training and even basic language instruction. Decision-making and discrimination abilities were the primary skills employed

Figure 5.3: Basic Logic for Opponent's Responses with Randomizer

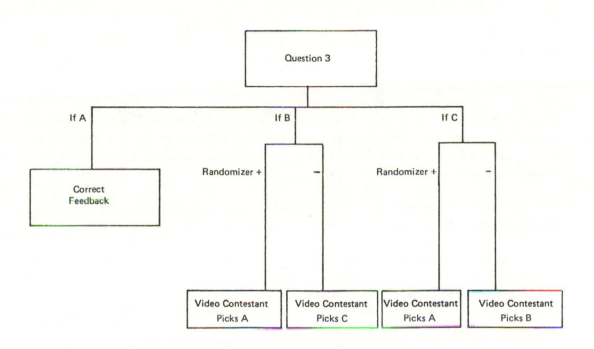

along with the occasional psychomotor activity which required quick reflex action. These all played major roles in successful interactive video adventure prototypes.

The Story

Whether an interactive adventure program has been written for training or purely for entertainment purposes, there is almost always a story underneath all the decision-making and discrimination activity that takes place. Saving the world from certain destruction is a popular plot. Usually in this scenario a maniacal band of villains is bent on taking over the world and destroying some wonderful part of humanity in the process. The mission of the game player, should he or she choose to accept it, is to stop the villains and save the world.

Other plots that seem to get endless service include:

- Rescue the victim
- Drive off the aliens

- Kill the monster
- Protect the helpless children
- Find the Holy Grail or other treasure
- Climb to the top

Designing an Adventure

With all interactive video adventures it is important to remember that the mission is secondary to the "business" of the adventure. It is the detail work that creates a sense of adventure and makes the mission worth doing. The archetypal TV adventure show, *Mission Impossible*, spent only about two minutes setting up the story line for each program, two minutes resolving the conflict and 50-plus minutes showing us neat little bits of business that took place during the adventure.

The Business

"Business" is an expression that is short for "stage business" or "screen business." It refers to the things that people do while they are waiting for major plot developments to occur. If the plot of an adventure is to find the Holy Grail, then the decision to begin the search sets up the premise, and the discovery of the Grail is the resolution. The rest is business. The following list describes some of the activities that would be considered business in a quest to find the Holy Grail or other such adventure:

- making preparations for departure
- choosing weapons
- choosing a path
- setting out
- crossing mountains
- crossing rivers
- crossing a desert
- crossing a wilderness
- finding a castle
- fighting off monsters
- getting through a labyrinth
- finding the secret chamber that contains the Holy Grail or treasure

In a spy adventure in which the player must decode secret plans to attack Cape Canaveral, finding out that there are plans sets up the premise. Turning over the decoded plans to the proper government agency is the resolution. The business could consist of the following activities:

- figuring out the location of the plans
- acquiring the plans

- breaking a code
- decoding the plans
- discovering a traitor
- keeping the plans away from the traitor

TURNING "BUSINESS" INTO INTERACTIVITY

"Business" includes all the activities that the viewer does for himself or herself. It is the mechanics of how the game is played. Once you decide what business you want, then you have to match it to the activities your system can perform. Table 5.1 matches key bits of business with well-known interactive video activities.

Table 5.1: Activities and Business

Activity	Business	How it Works
Multiple-Choice	Select weapons Select itinerary Select supplies Select anything	The computer assigns strengths based on choices. Later outcomes are based on these strengths.
Matching Fill-in-the-Blanks	Decoding Translating	A code is presented; use keyboard or joystick to enter translation.
Stay on the Path	Travel through an area Avoid pitfalls	Survival means staying within parameters. This activity is often created with computer graphic overlays or interleaving.
Racing	Chase	A joystick controls the scanning speed of the disc. A graphic overlay shows the relative position of adversaries during a chase.
Pathfinding Searching	Go through mazes Escape the Labyrinth Find an object	Graphic overlay of character or route over video still of maze or interleaving based on multi-level flow chart (up to five tracks). Object of search is on one track.
Shootouts	Kill or be killed	Graphic overlay of lasers or bullets, over target video. Branch to explosions and other activities.

Figure 5.4: Flowchart for Skateboard Odyssey Game

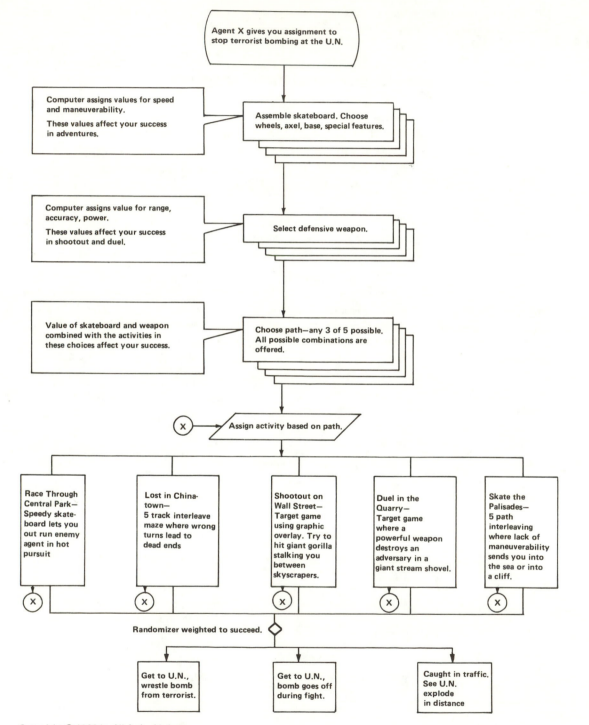

When the activities listed in Table 5.1 have been put together you will have the raw material of an interactive video adventure. Figure 5.4 does just that by showing the Level 1 (superstructure) flowchart for an interactive video adventure game called *Skateboard Odyssey*.

By taking a closer look at the Palisades segment of *Skateboard Odyssey* we can see how interleaving five different tracks allows the user to capture the true feeling of driving along a cliff beside the Hudson River.

If the skateboarder veers too far to the left, he or she will go over the side and into the river. If the skateboarder goes too far to the right, there is a steep cliff to run into. In between these extremes are thrilling little moments when the viewer heads toward the cliff but doesn't quite run into it.

To lay the sequence out with interleaving, the "straight down the middle of the road" track is put on every fifth frame starting with the third frame. Repeated crashes into the river go on every fifth frame starting with the first frame. The crashes are short and can be cut to in mid-crash to bring response time to absolute zero. Crashing into the cliff (as opposed to the river) starts on the fifth frame and is handled similarly.

We are now left with veering dangerously close to the river or the cliff in a scarey zig-zag motion. This effect is created by continuous motion sequences that are similar to the "straight down the middle" segments, but which provide the added excitement that is needed to keep the game interesting. These segments can be jumped to at any time, and are placed on tracks two and four respectively.

After a crash into the river or cliff the game ends and has to be restarted if the player wishes to begin again. Because a wrong move ends the game, the computer game control has to be fine tuned so that it is not too difficult (or too easy) to play. (Table 5.2 illustrates the interleaving process.)

Table 5.2:   Interleaving the "Palisades Sequence" For *Skateboard Odyssey*

Interleaved Tracks from Frame	Crash into river	Crash into river	Crash into river	Crash into river	Crash into river	Crash into river	Crash into river	Crash into river
From Frame 2	Veer toward river ⟶							
From Frame 3	Travel straight down the highway ⟶							
From Frame 4	Veer toward the side of the cliff ⟶							
From Frame 5	Crash into cliff	Crash into cliff	Crash into cliff	Crash into cliff	Crash into cliff	Crash into cliff	Crash into cliff	Crash into cliff

Given a producer who is willing to make the commitment and is given the equipment that can support it, this *is* the next phase in interactive videodisc programming.

SOAP OPERAS

Soap operas on commercial television seem to have become very sexy over the past few years. In the industrial business what we have are nonsexy soap operas. The mechanics of the presentation are the same, however. We will leave the quantity, quality and temperature of the sexiness in interactive soap operas to the imagination of the producers of interactive entertainment. But without making any moral judgments, we can say that there are plenty of people in high places in the home video distribution business who are convinced that the business got started because people wanted to see "X"-rated movies in their homes. All other forms of home video came later. Interactive video may or may not follow that pattern.

Anyway, "G"-rated soap operas have been ingredients in industrial video for years. They made their way into interactive video almost immediately. *People Skills,* the well-known Bank of America interactive video teller training program, is a cross in style between a soap opera and *Saturday Night Live.* The program starts with a heartbroken husband (Allen) telling the audience in a dramatic monolog how his wife (Teri) left him that morning because of their financial problems. Teri works as a teller in a local branch of the Bank of America. Half way through the program he appears in her branch to "see her and make things right again."

The disagreement between Teri and Allen provides a backdrop to suggest that people who come to work with personal problems have to find a way to maintain their professional attitude on the job in spite of their problems.

Teri is eventually turned down for a loan that could have saved her from the financial crisis, but she is convinced by a loan officer that she should encourage Allen to seek financial counseling. Allen accepts, straightens out his finances and they live happily ever after.

Okay, this may not be spicy enough by today's standards, but it is a good example of interactivity. Figure 5.5 illustrates the flow of discussion between Teri and the banker. It allows the viewer to see just what kind of responses will occur if the conversation is directed in one way or another.

Future interactive entertainment might make the viewer even more of a participant in the story (i.e., she could be the other woman; he could be the other man). If you think some of those soap opera characters are hateful now, wait until you are interacting with them personally.

Making the viewer part of the cast is easy. It takes little more than having the writer factor another character into his or her storyline. The interactive mechanics

Figure 5.5: Portion of Loan Interview Sequence from *People Skills*

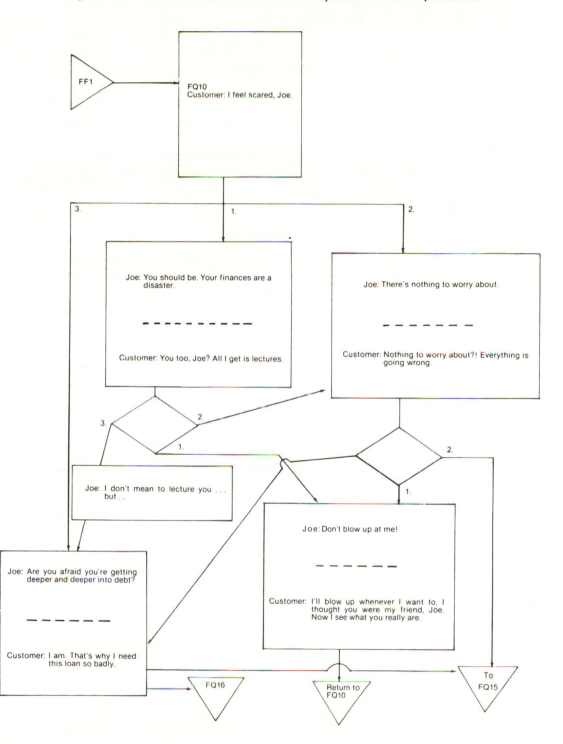

Note: *People Skills* is a teller training program prepared for the Bank of America. Reprinted with permission of Bank of America.

are nothing new, either. They've all been done before in industrial training soap operas.

CONCLUSION

We have looked into the future of home entertainment—a future that should have been here a long time ago. We know for a fact that the designs described above, as well as similar designs, are already on the drawing boards of America's great game companies. What we're waiting for is the right hardware system at the right price. We also need a daring entrepreneur who will invest the millions of dollars needed for software production. Unlike computer game production, interactive video cannot be produced by a creative hacker sitting alone in his basement. Interactive video is much more closely related to expensive movie-making.

While we are waiting for all this to happen, the research and development will continue in the industrial arena. Writers, producers and designers in major industrial media centers will continue to churn out innovative interactive programs. The result of their labors may far exceed the expectations of their corporate clients and enrich the lives of us all.

6 Interactive Video Database Applications

The archival capabilities of interactive videodiscs are legendary. What are not so legendary, and may be even more important, are the ways in which video images can be manipulated once they have been placed in a video database.

We will begin this section with a quick review of the archival capabilities of the videodisc and then move on to examine what can be done with video images once they are in place in a database.

THE VIDEO ARCHIVE

If you take a number of video images (somewhere between two and 52,000) and put them through a film chain, or copy them from a copy stand, you will have created a video archive. It is important to pay attention to field dominance during the copying process to eliminate frame flutter problems. If you do it right, you can lay down a series of still images that a viewer can step between, view or study.

The following is a semi-random list of the kinds of images that have already been transferred to videodisc:

Scientific Images

Photographs of space exploration
Astronomical observations
Geological images
Photomicrography

Medical Images

Radiographs
Diagnostic samples (e.g., tumors, lesions and blood samples)
Still photographs of proper medical procedures

Images for Travel and Tourism

Points of interest
Scenic attractions
Stores and hotels
Hotel rooms and facilities
Restaurants (decor and menus)
Maps

Collections

Famous paintings
Famous sculpture
Famous photographs
Historical artifacts
Memorabilia
Vintage automobiles
Trivia

Items for Sale

Catalog items
Fashions and accessories
Foods (from delivery or catering services)
Homes for sale
Tools and equipment
Parts and supplies

People and Services

Models (from a given modeling agency)
Prospective dates (from computer dating services)
Legal advisors
Consultants

Reference Books

Encyclopedias
Atlases
Dictionaries
Picture books

It has become clear from the lack of success many companies have had trying to market archival videodiscs that merely getting the images onto the disc is not enough. The question then becomes, "What do you do with a videodisc once the images are on it?" The answer involves the one thing few archival discs pay much attention to—the organization of the material on the disc.

why don't they ?

In this chapter we will review the possible organization schemes of video databases beginning with the simplest—the linear database. But, before we do that, it is important to note that all discs have the capability to *chapter*. What this means is that you can divide any disc into chapters or sections that can be accessed through a menu at the start of the disc. The following discussion concerns what happens after a specific chapter or section has been accessed.

THE SIMPLE LINEAR DATABASE

You have looked at the menu and selected a chapter. The particular chapter you have chosen consists of, say, 300 still frames lined up in a row. There are several local controls you can use at this point. These include simple step through, timed step through and play through. We will discuss each of these steps in detail below.

Simple Step Through (Forward or Backward)

As is the case with many commercial data discs, the viewer simply goes to the chapter containing the images he or she wishes to see and steps through them. Often, single-frame images are followed by frames that define or explain them. The rationale behind this procedure, of course, is to keep the images we are studying full screen so no text intrudes upon them. Text is then added in subsequent frames. To access those frames we hit a forward key (or whatever key controls our action) and click through the database one frame at a time, moving through the images either backward or forward.

Timed Step Through (Forward or Backward)

Timed step through provides minor variation to the simple step through technique. In this case, we add a timer to the controlling computer to advance the images at predetermined intervals. These intervals can range anywhere from 10 seconds to a minute. Again, the system can go backward as well as forward. This option can provide a good overview of the images available in a section.

Play Through (Backward or Forward)

Play through provides the opportunity to proceed through the images, backward or forward, at 30 frames per second (music can be added for the forward play).

Figure 6.1:   Typical Touch-Point Structure for Database·

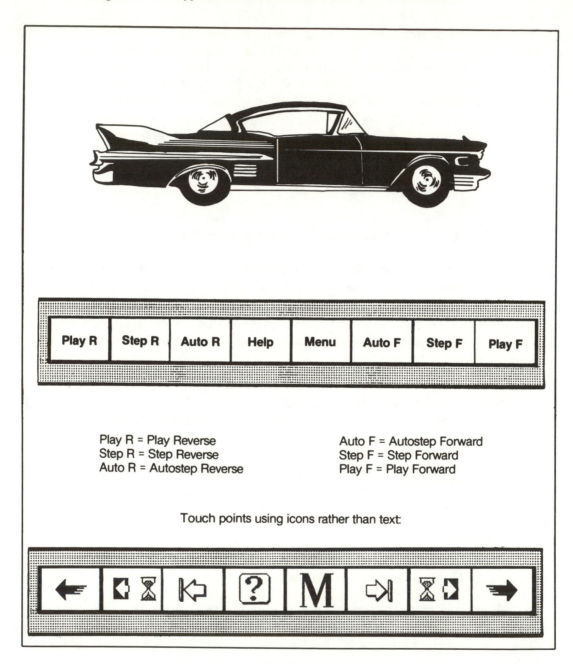

| Play R | Step R | Auto R | Help | Menu | Auto F | Step F | Play F |

Play R = Play Reverse   Auto F = Autostep Forward
Step R = Step Reverse   Step F = Step Forward
Auto R = Autostep Reverse   Play F = Play Forward

Touch points using icons rather than text:

The net effect is a very fast scan of the images, that can be entertaining in itself. The viewer has to stop the play procedure if he or she wants to study any of the frames. Play through can be considered as a minor diversionary mode of presentation that is more style than substance.

The local controls we have discussed above can be arranged as touch–points on the video screen. When added to a central menu they make up the most basic package of control options for a video database. A typical touch–point structure for such a database is shown in Figure 6.1. In this example, the function key for Help merely defines what each of the other function keys does. It might also explain how to get out of the program all together. Figure 6.2 illustrates a help frame for a linear database local control.

THE INDEXED DATABASE

Many databases are accessed through an alphabetized computer list that serves as an index. The indexed database works very well in two–screen systems in which a computer, with its own CRT screen, controls a video monitor that has images supplied by the videodisc player. When a segment of information is selected from the main menu, the computer screen displays a list of all the images contained within that particular database segment. A list might describe all the pre-Columbian art recorded on the disc or all the Packard Motor Cars shown from a vintage car collec-

Figure 6.2: Help Frame for Linear Database Local Control

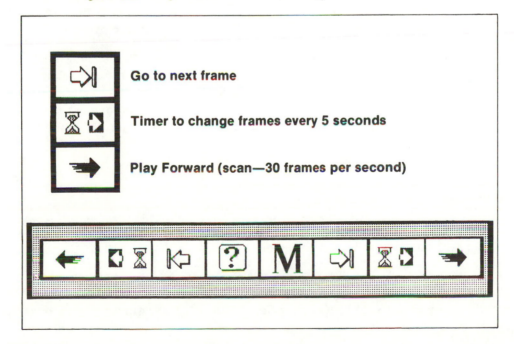

tion. There are a variety of ways in which the viewer can select an individual image from the database list. He or she can type a number into the computer keyboard control; use computer cursor keys to move an indicator arrow to the appropriate item on the list; touch a touch–point beside the appropriate item; or use a mouse to move an indicator to the appropriate item.

Cross-indexed databases, of course, merely provide the viewer with several different ways to get to the same data. If a viewer wanted to see an image of a 1950 Packard motor car, for example, he or she could find the image by accessing a menu of Packard motor cars, a menu of cars manufactured in 1950 or a menu of American luxury automobiles. The better the cross-indexing the more useful a database will be.

Local controls, such as those shown in Figure 6.1, could be available at individual frames to give the user greater control of the data. Perhaps the cardinal rule in the development of any database is that there can never be too many ways for the user to access information.

THE MOVING DATABASE

Motion sequences can also be included in the videodisc database. For example, when items in the database *move,* scenes of the item in action can (and often should) be made available to the database user. The motion sequences would be identified as such in the index. In our automotive database we could include short scenes of 1950s Packards driving down the road to go along with our still images of the car. The CAV videodisc allows for 30 minutes of full motion video playing time, and every second of motion takes up the space of 30 still frames. Yet, allowing for the sacrifice of some still frame disc space, the ability to add motion sequences to your database will probably seem like a valuable and attractive enhancement.

Since it is possible to record such a large number of still frames on a single disc, the best kind of database would include both still frames and motion sequences. If the motion sequences in the database are held to 20 minutes, there will be room for an additional 18,000 still frame images. This is probably a greater number of stills than most database makers are likely to acquire. So, don't hesitate to build motion sequences into your database.

Hypercard

Apple Computer's *Hypercard* may be the ultimate authoring tool for creating a database. Although it can be used for many applications, *Hypercard* works extremely well as a database manager.

In *Hypercard*, information is organized on computer images of cards which are generally arranged into stacks (stacks of cards). A unique feature of the cards is that they have buttons on them that can link them to other cards. There are buttons that go the next or previous card and there are buttons that are linked to different cards in different stacks. This feature allows for free association between many different cards in different areas and, in fact, begins to approximate the way the human mind jumps from idea to idea.

In an amazing moment of foresight Apple decided to allow the buttons that link cards to also send commands to a videodisc player. This means that cards can contain buttons that tell a disc player to go to a given video still frame or play a certain video sequence. So, any card in any *Hypercard* stack can be a video menu or the reference card associated with frames of video data.

Your database is now a stack of cards on a computer with information about sequences or still frames on a videodisc. The stack on outer space, accesses videodisc images of the moon and planets. Your stack or cardfile on great works of art can call up video stills from the videodisc on the National Gallery of Art. You flip from card to card and as you do, the disc image changes along with the cards.

As of this writing, *Hypercard* is a new product. Its potential has only begun to be tapped, yet, it is clearly a great tool for accessing and organizing a video database. Chapter 11 presents a detailed case study on how to use *Hypercard* to control sequences on a videodisc.

THE LIVING DATABASE

The structure of a video database should be functional and easy to get around in. This does not mean that it cannot be fun. Video databases, because of their use of images, lend themselves to structures taken directly from real life. This ability gives users a frame of reference to work with; a metaphor that can be more fun than the simple storage file images used by most computer databases.

The famous MIT *Aspen Map* project is the perfect example of a "living" database. It offered video access to all the streets in Aspen, CO at different times of the year.

The structure of the program made it easy to access that different information. Instead of asking your computer to show you "B" street in autumn, you could go to that street in autumn by moving down the other streets of Aspen until you got to it. If you wanted to see a particular building on "B" street, you could travel down the city streets until you reached it and then request to enter.

Movie maps like the *Aspen Map* project are excellent examples of the kind of creative elements that can be added to video databases. Other examples include shopping systems organized as shopping malls through which the viewer can move. In such instances the viewer can actually walk through the video images of buildings in a shopping mall and enter a given store before settling down to page through the video catalog of the store's products.

THE DIAGNOSTIC INTERACTIVE DATABASE

To some degree all databases are interactive, but how about a database that is so interactive you don't have to know exactly what you want before you access it? The diagnostic interactive database will ask *you* a few questions and direct *you* to the data *you* need.

One way to get to the data is to have the computer recommend something or someone by responding to specific input. For example, a video producer needs a Black male actor, aged 35 to 50, with a deep resonant voice, trim physique, and experience in comedy, drama and Shakespearean theater. A talent agency could have an interactive video system that would cross-check the required criteria to come up with the picture of an actor who might suit the part. The system could also include a few clips of the actor doing comedy, drama and Shakespeare. In the ideal case, the producer would select the actor by responding to a series of questions from the computer. The computer diagnoses the problem and responds by selecting and showing videos of possible candidates.

The sales advantages of such diagnostic interactive databases are discussed in more detail in Chapter 8. However, even outside the realm of interactive shopping, diagnostic interactivity has to be considered by any serious designer of these systems. The concept works for theater and film props, parts for cars, hotel rooms in Hawaii, products, gifts, suppliers, materials, plans, medical information, etc.

THE BUILD-IT YOURSELF DATABASE

What if, just for the fun of it, we came up with a database that had all the ingredients of a great game. Let's pick a sword and sorcery game. You get a video-disc with assorted dungeons, cavernous rooms, villians, heroes, damsels in distress, woods to walk through, rivers to cross and mountain ranges to scale. You can access them from a source list if you want to—to study them—but the real challenge is to turn them into a real, *playable* game.

Assembling components from a database of game parts could bring a new dimension to database usage; a database for the fun of it.

CONCLUSION

This chapter has explored the organizational realms of interactive video data-bases. Our intention has been to emphasize that there is more to building video data-bases than merely putting images onto a disc. The secret is in the architecture of the database—the structure of the museum in which you choose to house your collection. If you build your database with attention and dedication, and maybe a bit of the dramatic, you will end up with a database that realizes the highest objective for all databases—it will be used.

7 Interactive Video Information Centers

Interactive video information centers are stand-alone systems designed to provide quick and easy answers to questions from passers-by. Questions are often asked by means of easy-to-understand selections called *menus*. Answers often contain computer text, but also take advantage of the presentation power of videodisc technology (color, sound and motion).

Interactive video information centers are distinguished from point-of-sale (POS) systems in that they are intended to disseminate information, not *sell* products. While this difference may seem obvious, there are major considerations to be made in the design of each program structure.

The structure of an interactive sales situation can be thought of as an inverted pyramid in which all interactions funnel the viewer to a single action—a purchase. The interactive information dissemination structure, on the other hand, can be considered as a series of opening doorways that lead to more doorways. In such a structure viewers' options expand and movement through the material becomes more and more flexible. Figure 7.1 suggests the comparison between interactive video information and point-of-sale system structures.

Even though the structures of informational and point-of-sale programs are exact opposites, the two applications share similar requirements. Both for example, must be easy to use. Their structures should also not get in the way of the user's ability to accomplish the task at hand. This point becomes especially apparent when we consider the overall configuration of the hardware used.

**Figure 7.1: Interactive Video Information Center Structure
Versus Point-of-Sale System Structure**

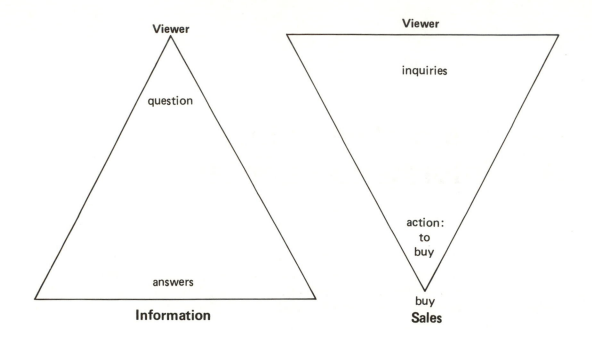

OVERALL HARDWARE CONFIGURATION OF THE
INTERACTIVE VIDEO INFORMATION CENTER

While there are a variety of possible configurations for interactive video information centers, there are certain elements common to all. Systems can be countertop, free-standing kiosk, through-the-wall units or desktop, yet each requires a clearly viewable screen dedicated to one or two viewers with indexing and menu systems that must be self-evident (as they would be on a touchscreen system). Global keys (Quit, Main Menu, Help, Step Forward, Step Back and Pause) should also be available to users. Variations, such as Directory instead of Main Menu or Introduction rather than Help, can be made to suit the vocabulary of the target population. General functions should be the same, however. The five or six function keys need not be part of the screen but can be located outside the video screen area through membrane keys, buttons, joysticks or whatever device seems appropriate for the individual use.

COMPONENT PARTS

We now have an interactive video information system consisting of a touchscreen with global controls that are located outside the screen area. The disc player and the CPU can be hidden behind a counter, in a separate control module or under the housing of a free-standing kiosk. The system may also need a data entry mechanism—such as a keyboard—so that information including names and addresses can be entered into the system.

SIGNS

Signs comprise another important part of an interactive video information system. They tell the user what to do. It is important, however, not to clutter the area with signs that end up adding confusion. On the other hand, past experience has taught us that the total absence of signs can leave the user without a clue as to what to do.

There are perhaps five or six basic steps that are vital to explain effective use. These steps should be listed on a sign on the face of the system and should be accompanied by a clear-cut statement of what the system is supposed to do.

The statement, "Everything going on in San Francisco," offers a nice description of what a San Francisco-based information kiosk system could tell you. The statement provides a good start for informing the public of your purpose. There are two additional items, however, that can enhance your hardware and signs; they are the attract mode and a live demonstrator.

THE ATTRACT MODE

The attract mode is a video segment designed to get the attention of passers-by. It invites them to come up to the information center and explains a little about how the system works. Due to the unique advantages of the videodisc player, the attract mode can be constructed so that it will play both backward and forward in an endless uninterrupted cycle. To accomplish this it is important *not* to include action that looks different when it is played backward. (It is amazing how many activities look the same going backward as they do going forward.) As long as people don't walk, things don't fall, cars don't drive and rivers don't flow, any repetitive motion looks great going backward as well as going forward. People talking (if you can't hear what they are saying), people dancing, people at work, special effects, static scenes, product shots and flashing graphics all work just as well backward as they do forward, and can be assembled into an endless attract mode.

Words which flash across the screen and are included in the attract mode should convey your system's benefits and main operational features. Phrases could include the following:

Touch Here for San Francisco Information
Current Film Reviews
500 Top San Francisco Menus

Critic's Choice of the Best Hotels
Points of Interest
Maps
Hours of Feature Attractions
Much Much More

Ongoing cycles of messages such as those listed above will help to get your information system used, especially if the video action includes sample shots of many of the items involved.

The attract mode is easy to create because most of the raw material for its construction will already be imbedded in your show. An even more indispensable item for the interactive video information system, and one too often overlooked, is a live facilitator.

THE LIVE FACILITATOR

Every point-of-sale or information dissemination system that we have ever been involved with has been improved immeasurably by the inclusion of live human beings who supported and explained the operation of the system. The need for human facilitators should be considered temporary—their support will inevitably need to be provided until free-standing interactive video systems achieve greater acceptance with the general populace.

It should be remembered that live facilitators supported the introduction of automated bank tellers until people began to feel comfortable with the machines. That acceptance *seemed* to come quickly, however, the fact is that automated tellers languished on the outer limits of banking for eight or nine years before they were generally accepted.

Interactive video information dissemination systems, not to mention video shopping systems, may benefit from the groundwork laid by the automated bank teller systems in acclimating the public to interfacing with machines. Live facilitators standing by to offer guidance and instruction will do even more.

How the Facilitator Operates

Attractive, well-dressed and well-trained personnel will stand beside the video terminal to invite the participation of passers-by. If desired, they can offer incentives to first-time users. For some of the more intricate systems, facilitators can operate the system for the customer. Usually, however, facilitators will offer only first-time demonstrations to be followed by guided discovery on the part of users.

A "scripting" technique developed at the University of Disneyland and employed at the EPCOT Center, required facilitators to memorize a scripted message. The message was complete with clever remarks and jokes. Our own experience with unprepared facilitators suggests that, while the system benefits greatly from the support of a well-trained facilitator, the involvement of an unprepared facilitator may actually hinder the acceptance of the system.

A unique feature of the EPCOT interactive video information centers was the inclusion of facilitators who were available via live video access. In other words, among the choices offered to users of the video information systems at EPCOT Center, was the chance to speak directly, via closed circuit TV, with a live "information specialist." The inclusion of this feature proved extremely popular, not only because the facilitator could offer up-to-the-minute status on events and occurrences around the park, but because specific advice on the use of the system could be offered as well.

TRANSPARENT ARCHITECTURE

In spite of the value of the live facilitator in an interactive video information system, there is no doubt that all such systems should be designed as though there were no facilitators there at all. Interactive video systems should have a structure that does not get in the way of the interaction. To borrow a phrase from computer programmers, the system should be "transparent." What we mean here is that the action to be taken by the viewer should be obvious every step of the way. As we have said, this begins with an attract mode that states the purpose of the system and tells, in very simple statements, how to use it. The same principle should be applied to menus.

MENUS

When designing menus for interactive information systems, special attention should be paid to the selection of the categories that the information is divided into. If the first menu a user encounters features six categories of information and there is a good deal of overlap in those categories, the user may get into trouble and just give up in immediate frustration. For example, if a user required information about a particular musical play and encountered the menu shown in Figure 7.2, what category should he or she choose? Is a musical play categorized under nightlife, entertainment, events, things to do or a place to go? It could be considered anything except, perhaps, sports. However, if the user looked for the play under nightlife and didn't find it there, he or she might go off in disgust. As silly as this may seem, many designers of interactive systems build menus with overlapping categories for a very practical reason—they are *selling* advertising space on the disc and they sell it by category. If they have six nebulous categories they can sell space to advertisers in all six different categories. Of course, when an advertiser only wants to buy space in one nebulous category he leaves the customer high and dry. Also, no one may get past the menu to see the ad, but such thoughts may not occur to our entrepreneurs.

Figure 7.2:   Example of a First Menu with Overlapping Categories

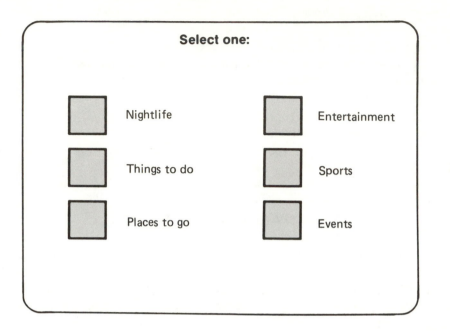

There are other things wrong with the menu in Figure 7.2. It does not tell the user what kind of information is being offered, nor does it tell the user what to do. It has been argued that menus are self-explanatory by their very nature, but extensive testing of signs done at ByVideo has indicated that *nothing* is self-explanatory to the majority of users. A well-designed information menu will be crystal clear (transparent) if it gives the uninitiated all the information that is needed without insulting the intelligence of the experienced user. Figure 7.3 illustrates a much closer approximation of what this menu should be like.

Sub-category Menus

It should be clear that a viewer's options can easily expand from a well-designed first menu. The restaurant category in Figure 7.3, for example, could lead to a sub-category menu, but here the plot thickens. While the rule about main menus is to make them as clear and simple as possible, rules about sub-category menus are far more difficult to pin down. You can give the obvious subdivision of the category or, if there are many different ways to subdivide the category, you can offer the user a choice. To continue with the restaurant example, you could ask the viewer to make the following decision—Do you want restaurants classified primarily according to price, location or type of food?

Designers must decide whether they want to give users unlimited choices at the sub-category level, or if they want to avoid the confusion that comes when users are offered too many choices or too many decision levels before they get to the information they seek. In designing information programs our rule of thumb is: Don't offer users more than three choices before they get information.

Figure 7.4 shows one path of a flowchart of an information dissemination system about restaurants that begins with a main menu as described in Figure 7.3 and then leads to a specific "high priced" restaurant. Notice the number of decision levels that the user has to go through before he or she gets to the actual information module. This may be too many and it might be better for information organizers to "bite the bullet" and pick one way to organize their information, at least for those users coming at the information from the start of the program. People who have been in the system for a while might be offered more complex menus with more discrete choices through a separate set of branches.

SETTING UP THE PROGRAM

Control Strips

Even though there are still a few interactive designers who fill the interactive video screen with control keys, buttons and touch-points, it has become common to

Figure 7.3:   A Well-Designed First Menu

SAN FRANCISCO TOURIST INFORMATION

Movies and Theater   Scenic Attractions

Museums and Parks   Shopping

Restaurants   Sports

Touch the category you want to see.

Figure 7.4:   Flowchart of an Information System About Restaurants.

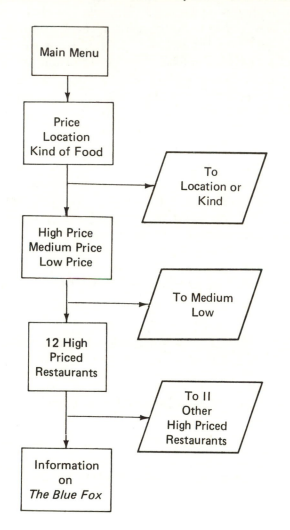

find these controls isolated to a single strip that can be found along the top, side or bottom of the screen.

The "control strip" usually consists of touch-points that are rendered to look like pushbuttons complete with three–dimensional shading. The number of touch-points on the control strip can vary from four to eight with certain very common touch-points becoming more and more standard as they appear indispensable to the operation of interactive information systems.

The indispensable controls on the control strip generally appear to be those that move the user within the immediate space of an individual segment or between major blocks of information. The controls that move the user through the immediate space of the segment are called "navigational controls"; those that move the user between major blocks of material are called "global keys."

Navigational Controls

Navigational controls operate on the premise that the user has to be able to move forward and backward through informational frames, as well as have the ability to jump out of any part of a particular segment to the end of that segment or the beginning of the next.

If you relate your video information system to a reference book, the typical navigational control might say "next page" or "previous page." It may also offer the ability to jump to the end of the segment with controls such as "scan ahead" or "next topic."

Global Controls

Global keys take users completely out of the sequence they are in. They can offer to take users back to the main menu, or they can give users the ability to access a help menu that leads to a variety of instructional segments. Other global keys can allow users to "quit" the program (at which time it will return to the attract mode), or they can give users the ability to freeze or pause any action sequence.

The most useful control strip configuration might contain a combination of global and navigational controls that are arranged so that the order of the touch-points has some logical relationship with the position of the segments in the program. (Figure 7.5 illustrates such a control strip.)

Control Strip Icons

Designers of interactive information systems have been searching for a set of international graphic symbols (usually called icons) to convey control strip messages.

Figure 7.5:   Sample Control Strip.

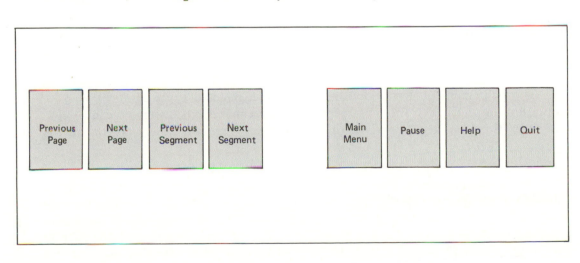

To use such icons or "road signs" of interactive video is always a great idea. Ideally, these messages require no translation and cross cultural barriers so they can be understood by people from a variety of backgrounds.

Apple Computer's *Hypercard* program allows each video control button to be identified with an icon. A good selection of icons is provided with the program. The selection is still not complete, however. There are still some functions that cannot be explained by a single icon.

So far, no one has been able to come up with the *perfect* universal set of icons. Traffic signs and international symbols work well in some situations, but not in others. The "stop sign" for example, seems to work well for "pause," but it could also mean "quit." Good luck to anyone with the talent to come up with such a system. In the meantime, using available icons combined with word buttons where no icons exist, seems to be the best way to go.

PUTTING IT ALL TOGETHER

We have looked at various elements of an information dissemination system. Now let's tie them all together in one example; a hypothetical travel information system.

Imagine a travel information system for a major city. The information centers are located throughout the metropolitan area—at the airports, major bus and mass-transit terminals, in the centers of tourist areas and in major hotels.

Since we (the authors) live in the greater San Francisco Bay area, we've designed this hypothetical system for San Francisco. Figure 7.6 shows the level one (most basic) flowchart for such a system.

The functions of the attract mode and main menu have been explained in the first part of this chapter. We have also touched on the function of help modes. Typical "Help" sequences can include one or more segments on how to operate the system, background on the company behind the system (to build user acceptance and trust) or an overview of the information organization (for those people who need the big picture).

The help menu is accessed from the main menu and each block of help information loops back to the help menu. The main menu can be accessed by a control strip function key. It is one of two or three function keys that are always present on the screen.

Category Introduction and Menus

From the main menu, the user accesses categories of information. The categories can, but certainly do not have to be introduced by a Category Introduction, which is

Figure 7.6:   Flowchart for Hypothetical San Francisco Travel Information System

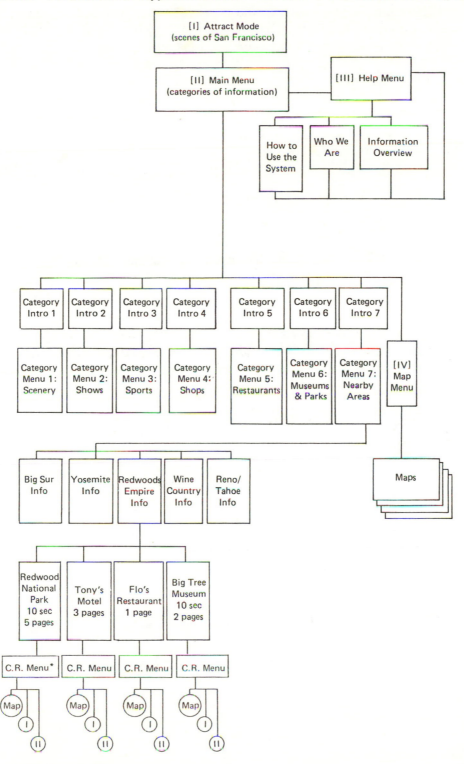

*"Cross Reference" Menu

10.

a brief motion sequence presenting the most impressive motion video on that category. These motion sequences should be visually dramatic and exciting, and no more than ten seconds in length. These sequences are part of the reason that inter-active video is used rather than simple computer graphics.

A Category Menu can use video images (assembled on an electric paintbox) to offer scenic representations of the information presented in the category. In other words, the category menu for Nearby Areas (see Figure 7.6), might show scenes of Big Sur or Yosemite National Park on a single frame. If preferred, icons could be used to represent these areas. (See Figure 7.7.)

Figure 7.7:   Icons for "Nearby Areas" Category Menu.

Yosemite National Park

Wine Country

Redwood Empire

Monterey and Big Sur

Gold Country

Text Information and Motion Sequences

The information is presented in the form of one to three single frames that combine an image of the location with text. If the information is relatively permanent, the text can be done as videographic text and included right on the videodisc. If, however, the text is in any way changeable, it should be presented as computer-generated text overlay that can be revised at a moment's notice.

There is room for a video motion sequence before the text information frames on each subject area (a short commercial, if you will). Two cardinal rules to be observed here are as follows:

1. The commercials should be no longer than 15 seconds (remember that we have a situation in which users are standing at a terminal requesting information);
2. The user should always be able to get out of the commercial (because even though the message is only 15 seconds long, it may seem too long to some people). A global escape key should therefore, always be available for use. We have interviewed users of information systems who tell us they felt "trapped in the commercial."

Finally, the information frames accessed by using the navigational control keys on the control strip can lead to a Cross-Reference Menu that allows the user to go to related subject areas or back to the Main Menu. Related subject areas could be information about subject areas similar to those that have just been reviewed by the user, or information about subject areas that are graphically adjacent to the area the user has just been looking at.

Maps

By taking advantage of the archival capabilities of the videodisc system, the user can be offered a variety of local maps in greater or lesser detail. Given adequate programming time, it is even possible to trace a path for the user to show him or her how to get to and from various locations. Questions most often asked might include:

What is the best route to get from San Francisco to Yosemite National Park?
What is the best route to get from this hotel to the freeway?
What is the best route to get from this kiosk to my hotel room?

The best way to answer these questions is to let the system draw a map.

CONCLUSION

This review of the capabilities and organization of interactive video information centers has focused on examples from the travel industry. Travel, however, is only one of the many applications to which these information dissemination systems can be applied. Museums and libraries have used them to guide people through their mazes of rooms and exhibits.

IBM's *Info Window* information system was said to be the most used attraction at the 1986 World's Fair. A variety of manufacturers have used interactive video to showcase complete lines of products at major trade shows. Hewlett-Packard, for one, designed a system to acquaint visitors to the Comdex Trade Show with the complete line of Hewlett-Packard Desktop Publishing, CAD-CAM and Computer Graphics products.

Interactive information dissemination is one of the most practical, interesting and useful applications of interactive systems. It grows even more powerful and becomes more economically viable as we add the ability to close a sale and take an order. We will discuss point-of-sale interactive systems in Chapter 8.

8 Interactive Video Point-of-Sale Systems

In Chapter 7 we briefly discussed the distinction between interactive video information systems and video point-of-sale (POS) systems. In that chapter we defined the structure of the video information system as a series of ever-opening doors leading the user to more choices and more information. The video point-of-sale system, on the other hand, was described as a series of fewer and fewer choices leading the viewer to a single action—a purchase.

Video point-of-sale systems can be divided into two major categories. These include "sell only" systems, which show off products and offer sales information but don't take the order, and "sell and buy" systems, which actually allow the shopper to place the order in the machine. However, since even "sell only" systems must get users to buy by referring them to live sales clerks, we will consider all point-of-sale systems to be "sell and buy" systems.

SALES FLOW

A major concern of professional sales people who deal with video shopping is sales flow. To complete a sale successfully, events must move smoothly from one stage to the next. The smoother the flow, the easier the sale. The worst thing that can happen to a sale is an interruption to the flow. Unfortunately, this occurs quite often with video shopping systems since mechanical or technical problems, or software difficulties created by faulty logic can easily interrupt the order-taking procedure, resulting in user frustration.

User Frustration

Assume that a shopper knows just what he or she wants to purchase on a video shopping system. If that shopper is unable to determine how to enter information to complete the sale there will be an interruption to the sales flow. This can occur at

83

any point during the sale, and if the user finds the video shopping system too cumbersome or frustrating to use he or she will most likely just give up. To avoid user frustration, designers of video shopping systems must make sure that the controls of the system are obvious and easy to understand, and that the information is organized in a fashion that is crystal clear.

KINDS OF VIDEO SHOPPERS

To determine how information should be organized to make it as clear as possible, designers must first take into consideration the different types of video shoppers. We have decided upon four different types: the first-time user, the browser, the person ready-to-buy, and the inquirer.

The First-Time User

The first-time user of a video shopping system does not usually want to get from the "start" to the "buy decision" as fast as possible. The first-time user wants to begin by learning something about the system. This might include a general discussion of the different products available on the system as well as a brief explanation of how the system works. The additional information required by the first-time user distinguishes him/her from the person ready-to-buy and the browser. Once the first-time user understands how the system works, however, he/she may then become interested in browsing through the products available on the system.

The Browser

The browser has needs that are different from those of the first-time user and the person ready-to-buy. The browser wants to explore the products by looking at several of them, studying them, learning about them and seeing them in action. The browser is the person to whom most of the space of interactive video shopping systems is devoted.

Since the browser does a great deal of bouncing around among products, it is important to remember that the goal of the video shopping system is to close a sale. If, in all the bouncing around, the sales flow gets interrupted, a potential sale may be lost. The organization required for the browser is different from that needed by the first-time shopper or the person ready-to-buy. Nevertheless, it too must be crystal clear.

Another important thing to remember about the browser is that he or she might not necessarily be a buyer, in which case it might be a good idea to insert some "close the sale" activities into the video presentation to turn the browser into a buyer.

The Person Ready-To-Buy

The person ready-to-buy is ready to make a purchase the moment he or she arrives at the terminal. For the person ready-to-buy, the process of browsing is superfluous and may even get in the way of closing the final sale. What is needed here is an organizational structure that gives immediate access to order entry, a "main line" from the start of the shopping process to the very end.

The Inquirer

As you see, several of these customer types can merge. The first-time user can become a browser, and a browser can become a person ready-to-buy. The inquirer is someone who needs guidance to be directed toward a particular product. He or she does not really want to browse around, but does not know what is available either. What is needed here is help. The mechanism for that guidance (the way the system helps the user determine what he or she wants) is present in the very best video shopping systems and may be nothing more than a series of questions about demographic data that zero in on a specific category of items. While this technique may seem obvious, it still must be organizationally transparent and easy to use.

RULES FOR THE CREATION OF VIDEO SHOPPING SYSTEMS

Now that we have looked at the different types of point-of-sale systems, the requirements of a continuous sales flow and the various kinds of shoppers, it may be possible to summarize what we have discussed in a few rules:

1. Make sure that the controls of the interactive video shopping system are obvious and easy to understand.
2. Make sure that the organization of the system is crystal clear.
3. Provide a path by which first-time users can get the background information they need to feel comfortable using the system.
4. Allow the browser an easy method to explore, learn about and look at products in action.
5. Build in a method for closing a sale, even with browsers who want to do nothing more than "look around."
6. Offer a method by which the inquirer can be guided to a product by asking a series of questions.
7. Allow the person ready-to-buy a method to get right to the buy decision in the shortest possible number of steps.
8. Devise an order entry mechanism that will not get in the way of the sales flow.

Controls

Methods for creating video control strips have been reviewed in Chapter 7. In many ways the control strip needed for a video shopping system would be similar to

Figure 8.1:   Control Strip for a Video Shopping System

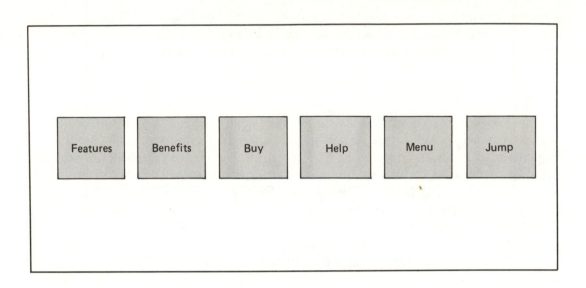

that needed for a video information system. One or two additional keys might provide all the added control required. Figure 8.1 shows a control strip for a video shopping system. "Feature" and "Benefit" buttons allow the user to access pages of information about the product. A "Menu" key accesses the previous menu. "Help" calls up instructions specifically tailored for that frame, and "Buy" goes to an Order Entry Sequence.

Organization

The crystal clear organization of the shopping system can be accomplished by two elements: the proper construction of menus, and the use of a "Welcome Sequence."

The Welcome Sequence

The Welcome Sequence is a short presentation that is played when a user begins to use the program. In it, a spokesperson explains, in 30 seconds or less, the overall organization and operation of the system. Since we have already added a "Jump Ahead" key to our control strip, anyone who does not want to spend the 30 seconds to listen to the welcome message can just jump ahead to the menu right after the sequence. Figure 8.2 presents the script for a welcome sequence written for a Hewlett-Packard point-of-sale program called *The Product Information Center*.

Figure 8.2:   Script for Welcome Sequence

Narrator: (walking into frame)

Welcome to the exciting world of Hewlett-Packard Business
Solutions. We've organized our program into three major
applications: desktop publishing, business graphics and PC
CAD. There's also information available on Hewlett-Packard
compatability, service and quality. It's all available to see
right now. Just touch the screen to get started, and, if you
want to leave any part of the program, just touch again and
you'll return to the previous menu. Why not make a
selection right now, and enjoy yourself.

Menu Construction

Menu construction is a difficult business. Each menu must contain a clear
presentation of all choices plus instruction about any action that must be taken to
make the menu work. Phrases such as "touch choice" or "touch one" should appear
early in the program to ensure that the user knows what to do.

It is also important that menus be looked at in series so that there is a logical
flow from one menu to the next. There will be times when people go from menu to
menu to menu. Figure 8.3 shows a series of menus from the Hewlett-Packard system.

There should be little or no redundancy in the names of the categories used in
the menu because this can lead to confusion. Also, enough instruction should be included on the menu so that the user will know what the menu is for and how to use
it. It is usually a good idea to have a narrator describe the use of each menu the first
time it is used. After that, a counter in the computer program can eliminate the
narration. Figure 8.4 matches a menu from the HP system with the narration used to
describe it.

PROVIDING A PATH TO THE SALE

The Direct Route to the Sale

Rules three through seven on our list of rules for the creation of video shopping
systems relate to the separate paths that should be created for the four different
kinds of interactive video shoppers. Since the most important customer is the
customer ready-to-buy, the best place to start is by creating a path by which that
customer can get to the buy decision as quickly as possible. Figure 8.5 shows such a
path.

Figure 8.3:   Series of Menus from Hewlett-Packard Product Information Center

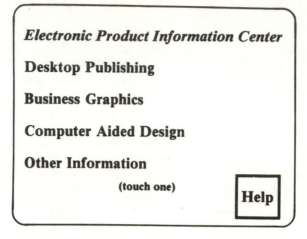

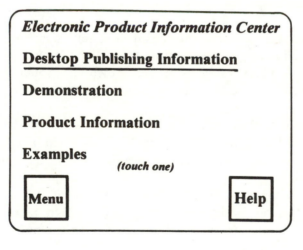

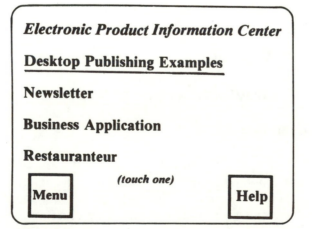

Reprinted with permission of Hewlett-Packard.

Figure 8.4:   Menu and Script

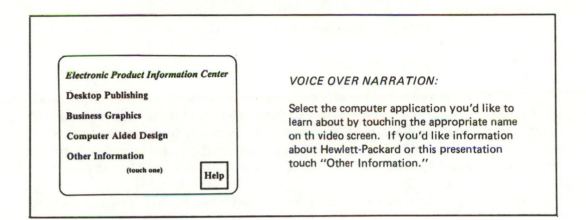

Reprinted with permission of Hewlett-Packard.

Figure 8.5:   Flowchart for the Person Ready-to-Buy

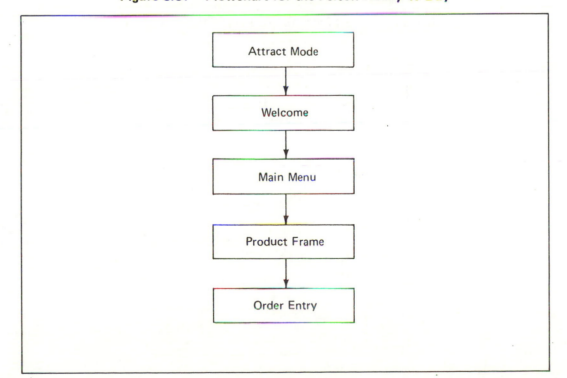

In spite of our best intentions, the sales path described in Figure 8.5 requires the customer to complete a sale by touching the screen at least four times and advancing through three menus. A faster way to accomplish this may be to place a buy key on the attract mode so that the customer can go directly to the order entry frame. A product list at the order entry frame can then clarify the item information as the customer is entering the order. Figure 8.6 shows the location of such a product list or index.

In the organization illustrated by Figure 8.6 the buyer jumps right to the order entry. At this point the buyer is either presented with an index, or types in the name of the product that he/she desires to purchase. The selection is then matched via computer software to the appropriate product information on the product list. This technique, which is similar to that used by computerized directory services, actually employs an "expert system" that could make the user's time very efficient.

You may be thinking that computerized directories require typing skills that you can't expect the general public to master. In this case we can use a visual directory that scans by the user at a rapid rate so that the user can stop at the appropriate frame when he or she sees the required product. Provisions can be made for controls including scan back, slow scan and step forward or back. Scan controls could be operated via the touchscreen or designated control keys on the control strip or on the keyboard. Figure 8.7 diagrams the video scan directory.

Figure 8.6:   Short Path for the Person Ready-to-Buy

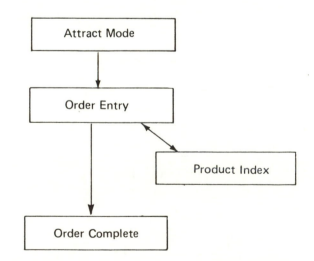

Figure 8.7:   The Video Scan Directory

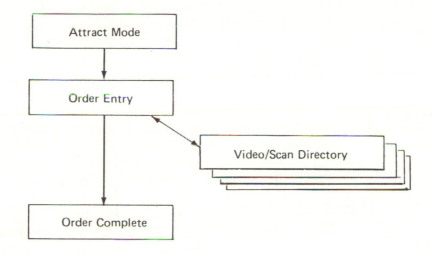

The "Additional Information" Block

Now that we have created the main trunk of the organization tree of our point-of-sale system and satisfied the needs of the customer ready-to-buy, it is necessary to deal with the requirements of the other customers—the first-time user, the browser and the inquirer.

The first-time user should have access to a special block of information that has been placed out of the main informational flow. This special information can be accessed by a global key on the control strip designated as "additional information" or "help." The key would access a first-time user's menu complete with detailed information explaining the operation of the system, an introduction to the sponsors of the system, an overview of the product selection, an explanation of how orders are processed and products delivered, as well as information concerning special offers and guarantees. Figure 8.8 illustrates the flowchart for this module of the point-of-sale system. Note that it incorporates the flowchart for the Video Scan Directory illustrated in Figure 8.7.

The Product Information Module

Our next step is to add the module on product information. This is the material with which the browser spends the most time. We will talk about the character of

Figure 8.8:   Flowchart of the Information Module for the First-time User

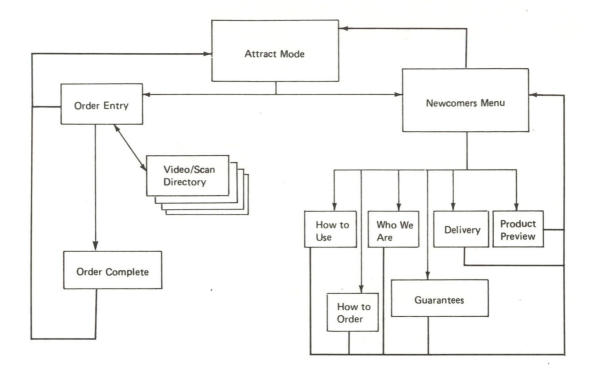

the video that goes into this module in a moment. First, let's consider how the information is organized.

Products should be divided into categories for easiest access. Information about each product can include a picture, some benefit statements, a list of features, technical specifications and pricing information. Perhaps the whole segment can begin with a brief motion sequence that shows the product in action. Figure 8.9 shows how all this can be put together.

It is likely that all the elements in the product A group could be separately accessed via control strip keys across the bottom of the screen. To make this easier to understand we can combine all the product A elements into a single segment designated as "Product A Information."

Figure 8.9:   Product Information Module

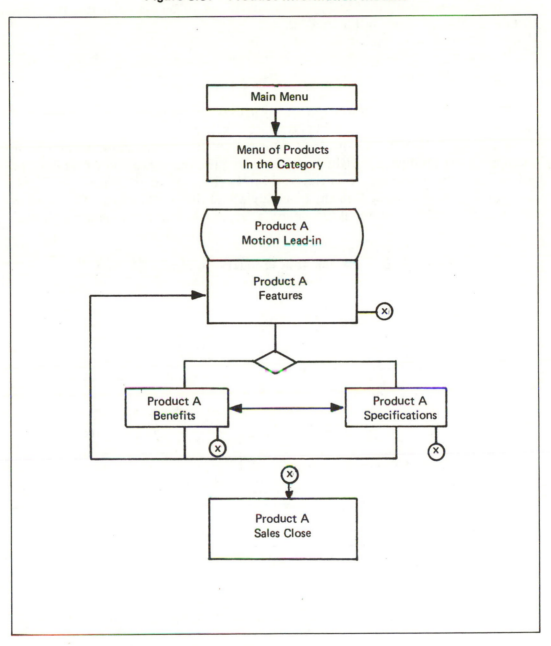

Logically, there is no reason why this segment of product information, ("Product Data Base" as some point-of-sale systems call it) cannot be placed before the order entry sequence, as shown in Figure 8.10.

The Inquirer Sequence

The video for the inquirer can be treated as a separate block of information generated either by the computer or by video text with information from the computer. As noted, this block is made up of questions that would allow the user to be directed to the product just right for him or her. The questions could lead into the product information sequence or (less efficiently) to a frame or frames created just for the "Inquirer Sequence." In any case they would fit into the overall flowchart as shown in Figure 8.11. Note that the variety of approaches to the material necessitate

Figure 8.10:   Placement of Product Information Module

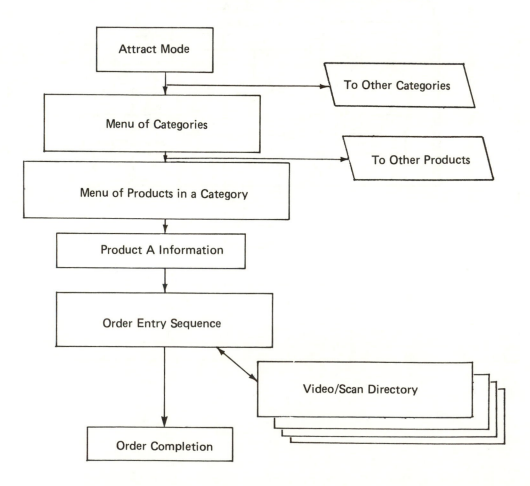

Figure 8.11:   Flowchart Including the Inquirer Sequence

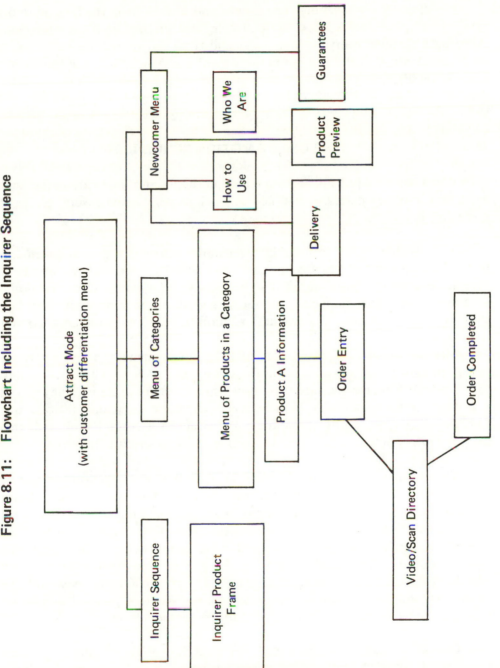

the creation of a special menu (or control strip) that identifies the customer type from the very start.

Needless to say, keeping clear menus and directions in the face of an organization so complex, seems like a difficult task. But the fact is, if the designers traced every path through every possible twist and turn, they could, in fact, come up with an organization that did not get in the way of the sales flow. . . an organization that was TRANSPARENT.

THE USE OF MOTION SEQUENCES

Where do you put motion sequences in the point-of-sale video program? There are a few obvious places, a few not so obvious places, and several places that are quite controversial.

The obvious places include the "welcome" sequence and several elements within the "newcomers" or first-time users sequence (i.e., who we are, how to use and product overview). (See Figure 8.4.) There is also good reason to use motion in brief sequences that show individual products at work. These sequences can range from short five-second shots of the product in action to full blown 60-second commercials.

The not so obvious places for video motion sequences in interactive point-of-sale programs are before each menu for a group of products. These menus have been called category menus because they represent a category of items. They can be given a stronger sense of identity if there is a brief motion sequence introducing them. Further, if the categories of products are arranged according to brand, then each category introduction can be a brief commercial for a specific brand.

The controversial aspect of motion in an interactive video point-of-sale program is the extent to which motion sequences are used. Many people argue that motion sequences use up valuable disc real estate and therefore limit the number of products that can be displayed on a disc. We do not agree. For example, if four frames are needed to display a single product and 5000 products are to be displayed on the disc, 20,000 frames would be needed. Since there are 30 minutes of motion or 54,000 still frames on a single disc side, using up 20,000 frames would still leave almost 15 minutes of motion available on the disc for product demonstrations, welcomes, newcomer orientations and category introductions.

One of the greatest advantages the videodisc offers producers of interactive point-of-sale systems is its ability to show products in full motion. To eliminate or minimize this benefit is wasteful.

Most designers of interactive video point-of-sale systems recognize this advantage and have included motion sequences in their programs, but this is where the controversy starts. The question is not whether video motion sequences should be included in the point-of-sale disc, but how long they should be.

Motion Sequence Length

After watching hundreds of video shopping customers at point-of-sale terminals, the answer begins to become quite clear—motion sequences should be as short as possible. They should be five, eight or at the most ten seconds long. Fifteen seconds is too long, 20 seconds is an eternity and 60 seconds will cost you your customer. He or she will simply walk away. It is important to remember that interactive video shoppers are not television viewers trapped in the safe confines of their easy chairs. They are *standing up* and walking around. Many of them are carrying shopping bags while others may have little children in tow. They are people in a hurry— they are going somewhere else.

To keep the interest of these hurried shoppers, video motion sequences should be kept short and sweet. Even though you could do a great commercial for a lawn mower, showing it in operation is enough; seeing a blender blend, or the real image of a mini TV picture on the video screen will do the trick. Each of these activities takes about five seconds and that is all you need.

Category introductions may take more time, about 15 or 20 seconds. A minute is still too long. The qualities needed to make a good category introduction are the same as those needed to make a good 15-second television commercial, especially the ability to convey a specific message in a compressed time frame.

The production people best suited to creating category introductions and most of the video motion sequences on interactive video point-of-sale discs are the people who know how to make short TV commercials. If you have limited time, go to them directly.

The controversy surrounding motion sequences in interactive video point-of-sale discs should be resolved. *Do* use motion for the welcome, newcomer orientation and product demonstrations, but keep these demos extremely short. Motion sequences used to introduce categories can be very effective but may be quite expensive and they, too, must be extremely short.

ORDER ENTRY SYSTEMS

Order entry systems finalize a sale by getting the customer's name, address and credit card information into the point-of-sale system. This enables the ordered product to be paid for and delivered to the right place.

One obvious solution to the order entry task is to have a live order taker on the scene. Most other methods of gathering order data are difficult and may result in a lost sale.

The true high-tech answer is to have the shopper's address and credit information on a credit card and a credit card reader attached to the video shopping system. The problem here is that, as of now, many credit cards do not carry this information (and if they do it cannot be read electronically off the card). The solution, of course, is to make your own credit card. When a customer applies for the card he or she agrees to make the information available to the system. You can imagine the massive marketing effort that would be required to initiate such a system, but a full blown interactive video shopping system may be worth it.

If the credit card carries address and billing data, the order entry problem is solved with a card swipe. The only other problem to be considered then, is the

Figure 8.12:   A Touchscreen Entry System

Select the appropriate letters to address your order:

A	B	C	D	E	F	G	H	I	J	K	L	M	N	O

P	Q	R	S	T	U	V	W	X	Y	Z	1	2	3	4

5	6	7	8	9	0	.	,	-

* *

Name:_____ Street Address:_____

City:_____State:_____Zip Code:_____

Message:_____

orders to be sent to addresses other than that listed on the credit card—in other words, gifts sent to other persons. A keyboard or touchscreen info-entering system may be the solution. As unpleasant as typewriter keyboards are to use, they may be the best choice. After all, they are a standard that the public recognizes. If you want to be truly flexible, offer an alternative—let the customer choose between a typewriter keyboard and a touchscreen entry system which arranges the letters in alphabetical order as shown in Figure 8.12.

The fill-in-the-blanks nature of the screen illustrated in Figure 8.12 is generally considered to be a great asset for obtaining proper information. It may be necessary, however, to explain how to get from one entry block to the next. Only a computer user would know, for example, that you have to tab to move from the "name" block to the "address" block.

CONCLUSION

In this chapter we have tried to show that the ideal video shopping system serves four very different kinds of shoppers and provides a flexible design that can accommodate any one of them. The ideal system is also dynamic enough to adapt to the shopper as he or she changes from one kind of shopping to another. There are fundamental rules that we have pointed out: the need for crystal clear organization, the importance of welcome sequences to get the customer started on the right foot, and menus that are complete, clear and easy to use. We have also presented program designs that offer background information, security and reassurance to the first-time user, an interesting product database for the browser and an "expert-system-type" question format for the inquirer who needs to be advised about products. We have also tried to suggest the need for a direct line to the order entry segment for those people who arrive at the video shopping system and are ready to buy.

Finally, we have looked at some of the more difficult questions of video shopping design: How much motion video is enough? How should order entry be handled? Should live sales people be part of the video shopping process?

Currently video shopping is one of the most researched and understood applications of interactive video. It is also one of most promising. We hope this review of our experiences in this area will lead to more and better video shopping systems.

9 Converting Linear Video to Interactive

As everyone knows, linear video can be converted into interactive video, but not without a great deal of cost and effort. What makes the conversion desirable is the fact that it takes considerably less cost and effort than it would to produce an original interactive video. But, the real problem with conversion is that the typical approach taken by most designers and producers results in programs that are rather poor examples of what interactive video is supposed to be.

In an educational or training video, for example, interactions should not be added to the program just to keep the learners awake. Ringing a loud gong every one or two minutes would have the same effect and perhaps be less painful for the learners.

Interactions should simulate the behavior that is being taught. And, to the extent that they zero in on the correct behavior—the behavior that grows out of the objectives of the program—they make interactive video better than linear video. If interactions do nothing more than interrupt the linear flow with irrelevant or obvious questions, they are, at best, a waste of money.

The question of how to convert linear video into interactive video, then, becomes very simple—what do you add and how do you add it?

In this chapter we will look at the programs that are the most likely to be converted from linear to interactive video. These include training and educational programs.

ELEMENTS THAT MAKE TRAINING INTERACTIVE

Learning Exercises

Of the elements to add to a linear program, the first (in importance but not necessarily in terms of screen time or production cost) are learning exercises.

The learning exercises you will add to your program depend on what you are trying to teach. If you are trying to teach students how to use a particular computer software program, then exercises in which the students hack away at a carefully prescribed set of activities are in order. If you are trying to teach students how to crosswire an RS232 cable, exercises in which students match one pin connection with a corresponding pin, will do the trick. If you are trying to teach bank tellers how to differentiate between stamps to use on certain types of checks, then matching the right check with the right stamp will be fine. The fact is, the video or computer screens that simulate these activities will have to be created anew, since they probably do not exist in the original video.

You will need to go through the linear video to figure out where the key information is presented. You will then need to find appropriate points to allow the students to practice the application of that information. It is important to pay special attention to the amount of information you present before you insert the learning exercises because you do not want to overload the students with data before you give them a chance to work with it.

Remediation

Once you have come up with the exercises, the next step is to remediate (tell the students whether their answers are right or wrong, and why). The creation of remediation, as already noted, is done best when it is tailored to specific exercises and not merely a repeat of some piece of a previous lesson that does not relate specifically to the exercise itself.

There are many different ways to present remediation. The best way is to show the result of a simulation. That is, if students are asked to crosswire a cable, one of the very best remediations would be to show them what happens when they plug in the cable they have wired. To make the remediation even better, an added commentary that explains how students went wrong with their crosswiring could be added. In this way, students can learn from their mistakes. In the example of the RS232 cable (with 25 pins at each end), the possible mistakes are numerous. This means that the remedial sequences will also be numerous.

The treatment of the remediation—the way in which it is presented—will depend on two things: First, the content of the program should be considered. If the program is a simulation of computer software, the remediation may lend itself well to being done on the computer screen. If the program is about interpersonal relations, you may have to show different kinds of people responding to statements from other people on-screen.

The second factor affecting the way remediation is presented is the *style* of the program, especially, how it relates to the way you decide to handle the housekeeping portion of the program. Often, the treatment of the housekeeping portions of the program will also provide a logical format for handling remediation.

Housekeeping

(For in-depth information about housekeeping in an interactive video program, please refer to *A Practical Guide to Interactive Video Design,* Knowledge Industry Publications, Inc.) For our limited purposes here, remember that housekeeping is the part of the program that handles the operational details. These include:

- How the program is run
- What controls are used
- How things are organized
- How the menus work
- What operational choices can be made
- What special features the program includes

Often, we recommend the use of an on-camera narrator and a didactic style for the housekeeping portion of the program. When this is done, the same narrator can easily be used to present much of the explanatory portion of the remediation. Other housekeeping devices such as a voice-over narrator with graphics, computer screens or text in a workbook are less effective because they lack the user friendliness of the on-camera person, especially if you have done your homework and found a person who is well suited to the demographics of your target audience.

PUTTING IT ALL TOGETHER

We have identified the elements of interactive video that need to be added to linear video to make it interactive:

- learning exercises
- remediation segments
- housekeeping sequences

There are several ways that these elements can be produced. You can try to match the setting with the production of the linear video by using the same narrator and location. Or, you can wrap the linear video with a framing device that introduces a new person (a tutorial type) in another setting such as a classroom, a TV control room, a computer room or a limbo setting (nothing but black). Another method would be to use a voice-over with titles that are revealed on the screen as the voice is speaking. (It would be dull but you could do it.) No matter how you choose to introduce the interactive sequences, the information should be presented clearly. The housekeeping segment should cover all the required operational details; the narrator should be suited to the demographics of the target audience; the exercises should accurately simulate the behavior you are trying to teach; and the "material" should be inserted at the correct spots within the linear sequences.

Table 9.1 lists the sequence of events in a linear video and the appropriate points for inserting housekeeping, exercises and remediation into the program.

Table 9.1:   The Organization of Pins on the RS232 Cable

Linear Video	New Interactive Sequences
Overview of RS232	Housekeeping (how to use this program—introductory menus)
	Menu (choice of lessons to learn first)
Ground Circuits	Exercises on ground circuits Remediation
Data Circuits	Exercises on data circuits Remediation
Control Circuits	Exercises on control circuits Remediation
Timing Circuits	Exercises on timing circuits Remediation
	Final Exam
	Closing Housekeeping

PRODUCING LINEAR AND INTERACTIVE VIDEO SIMULTANEOUSLY

To get from an interactive video to a linear video, take out the exercises, the remediation and the housekeeping. It is a subtraction process. Of course, there will be a few rough edges that will need to be smoothed out. You may not be able to get into or out of the demonstration segments of the interactive video cleanly enough to have them stand on their own as linear video. What you have to do in such cases is shoot additional transitional scenes. We are now implying that you recognize that you will want to "go linear" before you start producing your program. We are not talking about going back to existing interactive programs and revising them. At this point in the evolution of interactive programming, most people do not have a backlog of completed interactive programs for conversion.

Let's look at an example of how an interactive sequence can be replaced with a sequence that is linear. The housekeeping sequence that gives the program overview and explains the use of the menus and keyboard will have to come out, but a new, shorter "welcome" sequence may need to be added in which the narrator welcomes students to the program and gives a quick overview of what will be presented.

You may not want to scrap the exercises entirely. A linear program with exercises, after all, is better than a program with no student participation. You may want to lead up to the exercises with linear video, present a choice frame (or other kind of exercise frame), lay down 10 or 15 seconds of the frame and then present the right

Figure 9.1:   Original Interactive Program

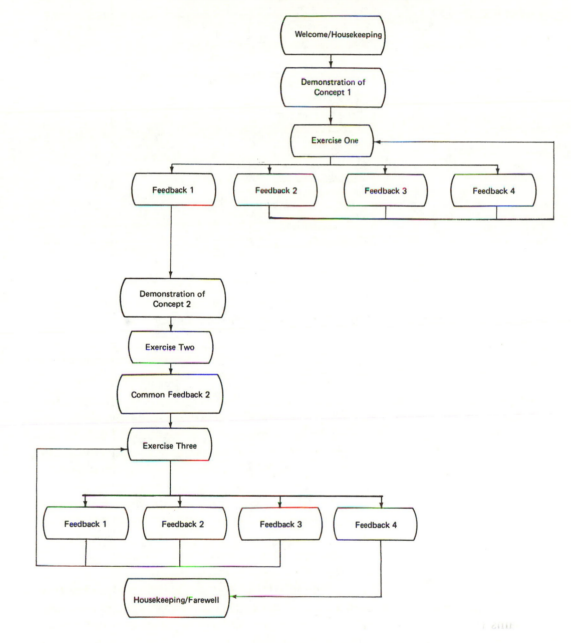

Figure 9.2:   Simplest Conversion to Linear

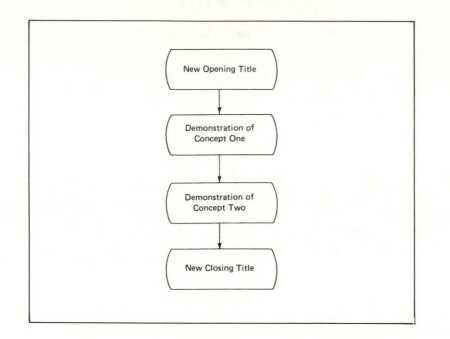

Figure 9.3:   More Typical Conversion to Linear

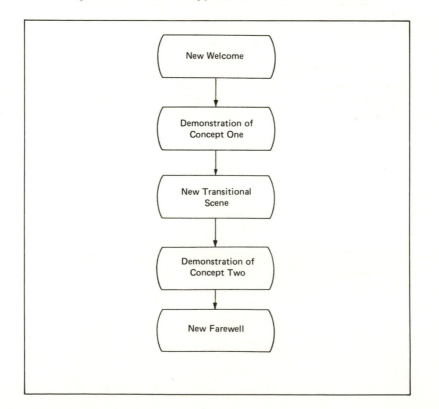

Figure 9.4:   Complex Conversion Retaining the Original Exercises

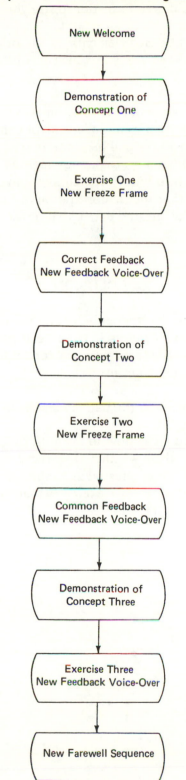

answer as feedback. This process may lack the dynamism of true interactive video, but educationally, we believe it is still superior to straight linear video.

Figure 9.1 shows the abstract elements of an original interactive program. Figures 9.2 and 9.3 illustrate two ways in which this program can be converted to linear video.

The important thing to remember is that, unless you have a perfect little jewel of a sequence that stands by itself as an independent demonstration (as is shown in Figure 9.2), it is usually best to try and incorporate some exercise material into the linear version. Figure 9.4 shows a very complex method with three different ways to turn interactive exercises into linear programming. In each case the required new scenes are noted. Also note that the Demonstration of Concept 2 was split into two demonstrations to match the greater detail that must have made two exercises necessary.

CONCLUSION

The economic picture for the production of any video program improves greatly if it can find multiple audiences and uses. If you can produce an interactive video program that can be recut into a linear program for those people who do not have interactive hardware, you stretch your investment.

If you can take your best linear program and make it interactive, you win again.

This chapter has tried to show how to do both of these things effectively. We have also tried to point out that, whether it is the process of adding material to make linear video interactive, or of substracting material to go linear, the secret of success is to focus on simulation. The exercises that you add (or modify) must simulate the behavior you are trying to teach.

10 Costs of Producing Interactive Video Programs

The cost of producing interactive video varies with the type of video that is being produced. Didactic approaches will cost less than dramatizations or documentaries—and no one-hour interactive video will ever cost as much as a 60-second prime time TV commercial. The best approach to determining the cost and time needed to produce an interactive video may be to begin by comparing it to the cost and time needed to produce a linear video of the same program. Then ask the question: How much more will it cost and how much longer will it take?

COMPARISON OF INTERACTIVE AND LINEAR PRODUCTION COSTS AND TIME

A video production can be divided into four standard segments that are used to determine cost and time. These include script development, pre-production, production and post-production. An added category for the special charges associated with interactive videodisc creation can be called replication and programming/authoring.

Table 10.1 compares the cost and time of a linear video with an interactive video that deals with the same subject. The sample program can be shot with several actors in a studio, a limbo set and very few props. The interactive production adds new graphics, additional scenes in limbo for remediation, plus graphics for menus. There is also more voiceover narration in the interactive video version.

Without going into the details of the samples used in Table 10.1, it is clear that they are fairly sophisticated productions, with large casts and crews. Nevertheless, the linear video can be shot in two days and the interactive video in three. The editing time is one and two days respectively. Offline editing is one day in either case. Graphics time is shown as one and two days (this may be spread out as parts of days over a longer period of time). In any case, there are more graphics needed in interactive video. The *special* interactive video charges (disc mastering and disc pro-

Table 10.1: Cost and Time Comparison of a Linear Video Versus a Similar Interactive Video

		LINEAR	INTERACTIVE
Scripting and Development:	Time:	Variable	Variable
Analysis and Design			$2,000
Scripting		$2,000	2,000
Approval and Revision			
Pre-Production:	Time:	2 Days	2 Days
(producer/director time totaled under post-production)			
Production Assistant		$500	$500
Video Production:	Time:	2 Days	3 Days
Facility		$4,000	$6,000
Crew		1,500	2,500
Equipment		1,000	1,500
Cast		2,000	3,000
Meals and Expenses		500	750
Tape Stock		500	750
Graphics Production:	Time:	1 Day	2 Days
Electric Paint Box		$1,000	$2,000
Character Generator		500	1,000
Graphic Artist/Designer		1,000	1,500
Audio Production:	Time:	1 Hour	1 Hour
Voice Over Narrator		$500	$500
Voice Over Recording		500	500
Music		500	500
Post-Production:	Time:	1 Day	2 Days
Off-Line Edit Facility		$ 500	$ 500
On-Line Edit Facility		2,000	4,000
Director		6,000	7,000
Producer		6,000	8,000
Interactive Charges:	Time:	n/a	1 Month
Disc Mastering			$2,000
Check Discs			750
Disc Replication			750
Programming/Authoring			3,500
Subtotal:		$30,500	$51,500
Contingency		$ 3,250	$ 5,600
Grand Total Linear:		$33,750	
Grand Total Interactive:			$57,100

Note: Dollar values are hypothetical and will vary according to the production and local prices for service.

n/a: not applicable

gramming) as well as the added time needed for instructional analysis and design are added factors increasing the cost of the disc.

The ideal situation would allow us to say that discs take one month more to make than linear videos and cost about 60% more. The fact is, it isn't as simple as that. That great unknown at the head end—scripting and development time—can stretch out for one project and not for another. An interactive video done for a client who has his or her act together can get done twice as fast as the linear video for the client who doesn't know what he or she wants.

SIMPLIFIED FORMULAS FOR COMPARING LINEAR AND INTERACTIVE VIDEO COSTS/TIMES

Since interactive programming costs more than linear, the only way to get an exact comparison would be to have the same people produce the same style of video in both linear and interactive formats. If they were equally efficient in each step, and as good at interactive as at linear, then the difference in time and cost would equal the additional cost and time of programming, replication and analysis shown as one month and $9,000 in Table 10.1. About 25% additional production costs would also be needed to shoot new scenes and to create new graphics.

Two simple formulas for calculating the additional cost and time of interactive programming over linear are presented below. (P.A.R. represents programming, analysis and replication costs, and p.a.r.t. stands for programming, analysis and replication time. It assumes some overlap in the areas of programming and replication time):

Interactive cost = Linear production cost + 25% + (P. A. R.)

Interactive time = Linear production time + 25% + (p. a. r. t.)

INTERACTIVE HARDWARE COSTS

In addition to the production and programming costs of interactive video, the cost of hardware should also be considered. A disc player can cost less than a good 3/4-inch videotape player, so any professional group that is considering installation of new video hardware and wants to go first cabin (3/4-inch), is not looking at an increase in presentation hardware if it decides to go interactive. On the other hand, each interactive video system requires a computer, so that cost must also be added in.

We could argue that most people already have computers that will work so, if they have to buy a tape player or disc player the costs would be the same, and there would be no big additional hardware costs in going to interactive video. However, if we do that we would be kidding ourselves.

Table 10.2:   Estimated Costs of Interactive Video Hardware

Disc Player	$2,000
Monitor	500
Interface Card	1,000
Cables and Connectors	100
Computer	1,500 to 4,000
Total Cost:	$5,100 to $7,600

Note: Estimates are deliberately nonspecific to allow for fluctuations in prices.

Most people already have a tape system in place which means that the disc system has to be added to it. A separate monitor may also have to be added. Computers that work well at driving disc players cost anywhere from $1,500 to $4,000, and then there is the question of text overlay. It costs from $1,000 to $2,000 for an interface board that will work. (See Table 10.2 for a rough price list.)

The costs are sobering, but the fact remains—if you want to attain the additional effectiveness of interactive video and add the new dimension that it brings to linear video, you have to pay the prices.

THE REAL COST COMPARISONS OF INTERACTIVE VIDEO

The only real way to assess the cost and effectiveness of interactive video is to weigh it against the things it replaces (i.e., hours of classroom time in the case of training, and whole stores—brick and mortar—in the case of point-of-sale systems). When you look at interactive video in this way it begins to make terrific economic sense.

Table 10.3 compares the training costs of a sales management course taught with stand-up instruction at a central location with the costs of training using decentralized interactive workstations.

Imagine the comparison between the cost of something like a shoe store building as opposed to an interactive video kiosk that also sells shoes. The cost difference is dramatic, and the comparative effectiveness of the methods of sale is readily available (perhaps hundreds versus tens of thousands of dollars). Inventory problems can also be simplified.

By Video, Inc. responded to just such a situation for Florsheim Shoes of Chicago. The average shoe store, it turns out, only stocks a limited number of shoe sizes. Florsheim makes a wide variety of shoe sizes but cannot make them available to every store. The interactive solution was to create an interactive catalog of shoes.

**Table 10.3: Comparing Interactive Video Training
and Stand-up Instruction**

Savings in Time:

Student time (in class)	10,000	(hours)
Student time (in transit)	2,000	
Instructor time	1,360	
Subtotal:	**13,360**	
Student time (interactive)	4,000	
Time Saved:	**9,360**	**(hours)**
	1,170	**(days)**

Savings in Dollars:

Student cost (class time)	$350,000
Student cost (travel time)	70,000
Student travel expenses	150,000
Instructor cost	4,760
Total cost of classroom instruction:	**$574,760**
Two interactive workstations at 20 locations @ $8,500	$340,000
First year savings	**$234,760**
(Yearly savings after hardware payoff)	**$574,760**

Note: Assume that a five-day course for 250 students a year has been replaced with two days of interactive instruction. Student cost (including fringe benefits) is $35 per hour. Travel time (to and from class) is one added day with travel cost of $600 per student. The class has two instructors and an average class size of 15. The class meets 17 times.

The customer was allowed to pick the style and color of shoe (most adults already know their shoe size) and then order the shoes from a central warehouse.

The immediate economic impact was improved availability of a wide variety of shoe sizes. The long-range impact could result in the savings of millions of dollars on the brick and mortar costs of new, fully equipped shoe stores.

CONCLUSION

The cost/time issues involved in interactive video are complex. There is the immediate added price of design, production, programming and added hardware. A straight comparison of interactive media and non-interactive media always makes interactivity look like a more expensive proposition. But, when you start comparing interactive training to the real alternative tools—centralized seminars and unmediated stand-up instruction—the cost comparison suddenly swings quite favorably in the direction of interactivity. The same thing happens when you look at video shopping versus the cost of bricks, mortar and inventory control. In the last analysis, the only realistic way to consider the value of interactivity is to weigh it against the alternate methods of doing the same job. When you do that, costs begin to look very favorable indeed.

11 The Computer's Role In Interactive Video

Instructional designers, video producers and other people starting out in interactive video often approach the computer component of interactive video with great trepidation.

Computer programmers, CBT designers and others who have a clear understanding of particular computer applications still need to know how the computer fits into interactive systems. But, more important, they need to learn how to use the computer effectively in interactive video.

Basically, the computer's job is very simple; it sends commands through a cable to the disc player. It tells the disc player what to do, and then gets a message back that says, "I've done it!" Of course, there are a variety of other things that the computer can do as well. It can create graphics that are inserted between video sequences or overlayed on top of them. It can tabulate scores or other data. It can ask questions and select video segments based on the answers. The way in which the computer is involved in the interactive system will greatly affect the kind of computer that is used.

KINDS OF INTERACTIVE CONFIGURATIONS

There are a variety of interactive video hardware configurations, each suited to a different application. Many of these configurations assign a unique role to the computer system. Table 11.1 shows several different kinds of interactive video programs with increasing computer involvement.

In the first configuration (video images only—1 screen) the computer is nothing more than a controller. The only screen is a video monitor that shows images provided by the disc player. The computer screen is either blank or preferably, it isn't

Table 11.1: Interactive Computer/Videodisc Interface Systems

Video Only Images			Video and Computer Images	
System 1	**System 2**	**System 3**	**System 4**	**System 5**
1 Screen	2 Separate screens	1 Screen switching between video and computer	1 Merged screen with computer overlay	Merged screen with high definition computer images

there at all. The purpose of the computer in this system is to act as a controller. When the interface mechanism of the computer (usually the keyboard) is replaced by a device that is unlike a conventional computer keyboard, a touchscreen for example, it becomes clear that the system's designers are trying to hide any resemblance the system might have to a conventional computer. Applications for such video systems are aimed at those users who don't want to know that a computer is involved in the process. Point-of-sale systems are the most common example of this configuration. One problem with this system, however, is that once the disc is mastered only information stored on the disc can be shown.

The second system listed in Table 11.1 has video and computer images but they are confined to separate screens. The computer's task here is a bit more complex than it was in the first configuration. In this system the computer must generate information for its own screen as well as control the action of the disc player. One big advantage of the two screen system is that no special circuit boards are needed to mix the computer and video images into a single image. A simple RS-232 connection can link the video disc player and the computer.

The third system (computer and video images on the same screen but not mixed) is easier to design and produce, and less expensive than the merged image system. And, depending on the computer used, it may not require a separate board. The only drawback of such a system is that you have to deal with one image at a time, which can make it less effective than the two screen system.

The fourth system is the merged video and computer image system. It is also the most common and useful for applications such as training. This configuration overlays text and graphics onto the video image. It can generate screens that block out parts of the picture and mask images that can be revealed at other times.

The fifth system does far more than overlay computer text and graphics over the video image. It uses the computer to create high definition "video quality" images and then integrates those images into the video presentation. In this way the computer image almost becomes part of the video image and can be manipulated far faster than even the very best video image. This kind of integration is used in high-level simulations including arcade video games. Space ships, for example, that must scoot over some extra-terrestrial plane, can do so with great speed and maneuverability. The interactive equipment needed to create such a system would place special demands on the disc player, the computer, the software used to create the program and on the programmer who does the actual creation. (See Table 11.2 for the components required in each of the five computer/disc interface systems discussed above.)

COMPUTER COMMANDS

How does the computer manipulate the videodisc system. There are actually only seven basic commands—all other functions of the disc player are variations of these seven. The first command is an audio command and, to some degree, it is

Table 11.2: Typical Computer/Videodisc Interface System Components

System	Disc Player	Computer	Monitor	Player/Computer Interface	Software	Author
1 video screen (no computer images)	industrial	PC or mini	video monitor	RS-232 cable	authoring tools or Basic, Pascal and player programming manual	limited experience
2 screens 1 video 1 computer	industrial	PC or mini	video monitor, and computer CRT	RS-232 cable	authoring tools, Basic or Pascal	limited experience
1 screen switching video and computer	industrial	PC	composite or CRT monitor	RS-232 cable	authoring tools, Basic or Pascal	some programming experience
1 screen video and computer merged	industrial	PC	composite or CRT monitor	RS-232 cable and overlay board	authoring system	limited experience
1 screen video and computer generated high-quality graphics	industrial	PC	composite or RGB monitor	overlay board, parallel (IEEE-488) interface or specially constructed interface	"C" or other high-level compiled language	experienced disc programmer

separate from the others. The remaining six commands are controls of the video (and the audio that goes with it). These seven commands are listed below:

- Command 1, Audio channel 1/2—on/off
- Command 2, Search to frame #_____
- Command 3, Play forward (or reverse) until frame #_____
- Command 4, Stillframe (freeze)
- Command 5, Step forward or reverse
- Command 6, Get status (frame #) from player
- Command 7, Jump to frame #____ (for players with "quick jump" capability)

Of course, the commands can be combined. For example, if commands two and three are combined the disc player is in effect told to:

"Search to frame _____ and play." until frame # _____.

More commands can be added so that the string includes commands two, three and a repeat of two and three. This will read:

"Go to frame 'X' and play to frame 'Y,' then go to frame 'Z' and play."

It does not take a very complex instructional design to ask for this kind of computer control, yet it requires something very demanding of the disc player. The disc player has to recognize frames "X," "Y" and "Z" when it gets to them, and it has to tell the computer that it is there.

Now, it is very easy to send commands from a computer through a cable to an instrument on the other end. But in interactive video it is necessary to get a message *back* from the disc player telling which frame is showing. Getting the current frame number is important so that computer actions can be synchronized to the exact frame showing at the time.

The process we are describing has several names—it can be called a closed loop system (as opposed to an open loop one-way no feedback system), or it can be referred to as "getting status back" from the disc player. It requires that the computer programming and both the disc player and computer used to control it be a bit more sophisticated than they might otherwise have to be.

Fortunately for those of us who are not programmers, authoring systems are being developed that have already put together most of the basic commands and the sophisticated protocol needed for this kind of computer control. They come prepackaged with neat little menus. So, very often, all we have to do is fill in the blanks and the authoring system does the rest.

AUTHORING SYSTEMS

Authoring systems are not the only way to get your disc programmed. If, for example, all your interactions involve the video screen and you have some experience with computer programming, the dealer who sold you your videodisc player can supply you with a set of instructions that will enable you to issue the basic videodisc commands to run most *super-simple* interactive videodisc programs.

At the other end of the spectrum are the *super-complex* applications like action "arcade-style" games, enormous databases or seamless simulations. Most authoring systems are not equipped to help the designer build the programs that run these applications. Even if an authoring system were used, a highly skilled programmer would probably be needed to expand the system. In the long run, it might actually cost less to have the project done by a professional programmer without the authoring system.

Fortunately for most of us who are trying to develop interactive video, most programs fall between the super-simple and super-complex—especially interactive video *training* programs. And these programs can be helped by a good authoring system.

A good authoring system should provide a clean, easy method for executing the basic commands with a minimum of keystroking or number crunching. It should also provide a framework for the CBT portions of training programs and provide a good structure by which the video elements can be integrated into the overall lesson. The authoring system should also aid in the design of graphic elements of the program, whether they appear on the computer screen or the video screen.

There has been a great deal of criticism of many authoring systems for training because they limit the interactive exercises from which the program designer can choose. Actually, some of the older interactive authoring systems did nothing more than ask simple true/false or multiple choice questions. When the answer was wrong the program simply looped back and played the section of the original video that presented the basic information—it was a weak instructional design that produced poor interactive video.

Today's better authoring systems not only offer the designer a wide range of instructional strategies, they also allow the programmer "hooks" so that he or she can get at the underlying control program without destroying the exterior structure. What this means is that the programmer can expand the capabilities of the authoring system by adding computer routines that permit specialized exercises or other activities.

The original distinction between an authoring system and an authoring language is that an authoring system can be expanded to suit the specialized needs of the developer.

Today, some designers prefer smaller subsets or parts of authoring systems called "authoring tools." These allow the designer to handle specific routines. The authoring tool can be combined with other tools or set aside depending on the project.

All in all, todays authoring systems can provide a good start for designers who are taking their first step into programming interactive video. If a powerful expandable authoring system has been selected, the system will be able to accommodate designs of increasing sophistication as the user's needs and creative ideas grow more sophisticated.

Figure 11.1 provides a list of many popular authoring systems.

SELECTING SOFTWARE AND HARDWARE

One of the most important things to understand when putting an interactive video system together is that many trade-offs must be made. In order to make these trade-offs correctly it is important to have clear measurable goals.

There are so many different ways to do things, each with its strong and weak points, that it helps to have all the particulars identified and set out in front of you before you begin. For example, the need to have the user's response occur within 1/10th of a second is important in programs that are trying to teach hand-eye co-ordination. To achieve this 1/10th of a second level of interchange a response that occurs on the computer screen or in the speakers may be required (i.e., keyboard clicks, computer beeps or sound effects). If the response must occur on-screen within 1/10th of a second, the disc geography might have to be more sophisticated or the control program may need to be written in assembly language or in a compiled language (these languages are faster). In other words, you may need to trade response time for programming time.

Another typical trade-off is memory size versus access time or computation time. For example, you can store 550 megabytes on a **CD-ROM** but getting the information may take two seconds. Having the information in main memory would make the information available within a billionth of a second, but main memory currently holds a megabyte at the most. The trade-off affects hardware and programming.

Another important consideration is who will be using the system. It probably will not be the system designer or programmer, so always keep *the user* in mind— will the operation of the hardware be so complicated or unusual that the user will end up worrying more about which button to push than about learning the desired material? It might be valuable to interview user groups before you design the system. Find out how the users think, and what they consider the most logical way of

Figure 11.1:   A Partial Listing of Authoring Packages

ADROIT (Applied Data Research, Route 206 and Orchard Road, Princeton NJ 08540, 201/874-9000).

AUTHORITY (Interactive Training Systems, 4 Cambridge Center, Cambridge, MA 02142, 617/497-6100).

COURSETEN (ForceTen Enterprises Inc., 3845 Dutch Village Road, Halifax NOV B3L 4H9 Canada, 902/453-0040).

CDS and CDS2 (Electronic Information Systems, 2630 East Fox Hunt Drive, Sandy UT 84092, 801/942-2260).

DDT (Digital Techniques Inc., 10 B Street, Burlington MA 01803, 617/273-3495).

THE EDUCATOR (Spectrum Training Systems, 18 Brown Street, Salem MA 01970, 617/741-1150).

GENESIS (Sony Corporation, One Sony Drive, Park Ridge NJ 07656, 201/930-6000).

IDEAS (OmniCom Associates, 407 Coddington Road, Ithaca NY 14850, 607/272-7700).

THE INSTRUCTOR (BCD Associates Inc., 205 Broadway Technical Center, 7510 North Broadway Extension, Oklahoma City OK 73116, 405/843-4574).

INSIGHT PC PLUS, INSIGHT 2000 PLUS, and *INSIGHT 70* (Whitney Educational Services, 415 South Eldorado, San Mateo CA 94402, 414/341-5818).

JAM DISC WRITER (JAM Inc., 300 Main Street, East Rochester NY 14445, 716/385-6740).

LASER WRITE (Optical Data Corporation/Video Vision Associates, 66 Hanover Road, Florham Park NJ 07932, 201/377-0302).

MENTOR/MACVIDEO and MACAUTHOR (Edudisc, 3501 Amanda, Nashville TN 37215, 615/269-9508).

MICROTICCIT (Hazeltine Corporation, 10800 Parkridge Boulevard, Reston VA 22091, 703/620-6800).

MCGRAW HILL INTERACTIVE AUTHORING SYSTEM (McGraw Hill Book Company, 1221 Avenue of the Americas, New York NY 10020, 212/512-2000).

PASS (Bell and Howell, 7100 North McCormick Road, Chicago IL 60645, 312/673-3300).

PHAROS (Interactive Video Concepts, PO Box 21, Media, PA 19063, 215/891-9080).

PILOT PLUS (Online Computer Systems, 20251 Century Boulevard, Germantown MD 20874, 301/428-3700).

PLATO (Control Data Corporation, 6003 Executive Boulevard, Rockville MD 20862, 301/486-8030).

QUEST (Allen Communication, 140 Wayside Plaza II, 5225 Wiley Post Way, Salt Lake City UT 84116, 801/537-7800).

SAM (Learncom, 215 First Street, Cambridge MA 02142, 617/576-3100).

SCRYPT (New Media Graphics, 279 Cambridge Street, Burlington MA 01803, 617/272-8844).

TENCORE (Computer Teaching Corporation, 1713 South Neil Street, Champaign IL 61820, 217/352-6363).

USE (Regency Systems, 3200 Farber Drive, PO Box 3578, Champaign IL 61821, 217/398-8067).

WISE (Wicat Systems, 1875 South State Street, PO Box 539, Orem UT 84057, 801/224-6400).

VAX PRODUCER (Digital Equipment Corporation, 12 Crosby Road, Bedford MA 01730, 617/276-1431).

VIDEO NOVA (Video Nova, Suite 500, 400 Renaissance Center, Detroit MI 48243, 313/861-6034).

Source: *The Videodisc Monitor,* March 1986 issue. Reprinted with permission.

getting things done. Build your hardware configuration around the user and choose an authoring system that will support *that* hardware configuration.

Designing a system that is similar to those with which the users are already familiar, will help them to adjust. Think of the desktop metaphor that is used heavily in office automation systems. There are metaphorical in-baskets, out-trays, work areas, folders; the works. Our "Dream Machine" controller (see Appendix at the end of this chapter), was also easier to use because we made it look like the remote control for a TV set. Using a metaphor is an excellent starting point for the development of an interactive system. The authoring system you choose should allow you to create such a device.

Metaphors, user tendencies, mechanical trade-offs, system flexibility and hooks for authoring upgrades are all tied together in Figure 11.2.It's a decision table to help you select an authoring system.

CRITERIA FOR MAKING HARDWARE AND SOFTWARE COMPARISONS

The interactive video market has only recently begun to stabilize. Level 2 training applications were center stage for a time, only to be supplanted by arcade game and point-of-sale applications. Now training applications have staged a return, arcade games are completely out and point-of-sale is maintaining a steady presence in the market. With each shift in interest the demands on the hardware, and in fact, the hardware configurations themselves keep changing.

What Does It All Mean?

The main point to be made here is that models have been changing. In their praiseworthy efforts to meet the changing demands of their customer base, manufacturers have continually modified their players. Fortunately there is compatibility within most individual product families and things have at least stabilized enough so that an investment in a laser videodisc player is an investment in a technology that seems to have figured out its capabilities and its purpose.

Designers of authoring systems, too, have finally found what instructional designers demand of them and many have answered those demands. As noted, Figure 11.2 presents a worksheet for assessing the capabilities of authoring systems. We don't need anything quite as elaborate as that to compare disc players since all systems do the same basic operations. Criteria, then, come down to a few simple, but very important items:

- Reliability—this is not as big an item as you may think it should be since the major manufacturers are all giant video companies. It's a little like asking, "What is more reliable a Ford or a Chevrolet?" Or, to be more specific, "Which car stereo is more reliable, Sony, Hitachi, Pioneer or Phillips?"

Figure 11.2: Authoring System Evaluation Form

Name: _____

Hardware Control:

Players [   ] Disc 1   [   ] Disc 2   [   ] Separate Audio
Controller [   ] Touchscreen   [   ] Keyboard   [   ] Mouse   [   ] Stick
Screen Control   [   ] Video Screen   [   ] Computer Screen
Status Back   [   ] Available

Function Control:

[   ] Separate Audio Channel 1 and 2 (off/on)
[   ] Search to
[   ] Play to
[   ] Stillframe (freeze)
[   ] Step Forward/Reverse
[   ] Get status from player
[   ] Jump to

Graphics Creation:

[   ] Character Generating
[   ] Pre-programmed Menus
[   ] Masking
[   ] Highlighting
[   ] Computer Graphics Creation
[   ] Video-Style Graphics (high detail)
[   ] Image Digitizing (photo-style imaging)

Pre-programmed Instructional Strategies:

[   ] Matching   [   ] Interrupt
[   ] Counting   [   ] Consequence remediation
[   ] Fill in the blanks   [   ] Drills
[   ] Type it in   [   ] Lightning round (timed action)
[   ] Multiple-choice   [   ] Game show
[   ] Spider webs   [   ] Optional analogies

Simulation Strategies:

[   ] Menu Driven Branching
[   ] Seamless Branching
[   ] Graphics Overlay Motion Control
[   ] Simulation of Computer Software and Its Use

Evaluation Capabilities:

[   ] Pre/Post Testing
[   ] Pre/Post Test Comparison
[   ] Exercise Scoring
[   ] Exercise Score Comparison and Evaluation
[   ] Exercise Item Evaluation
[   ] All-Student Comparative Evaluation

- Durability—Your choice of player may depend on your need for durability. If your system will run 24 hours a day in a public place, you may want to spend a little more to get a more durable player, one with an industrial spindle motor, for example.
- Service and Support—Here is a thornier issue. Some manufacturers have made a great effort to support their disc players once they are installed. That effort may vary geographically, so it is not always constant. It is critical to assess manufacturers by their willingness to provide service and support for the systems that they sell to you.
- Authoring System Support—Not all authoring systems will work with all players. If you are putting a system together to meet a near-term deadline, be sure that the authoring system you select will work with the disc player that you choose. If you have an authoring system that is just right for you, pick the disc player accordingly.
- Special Features—The *quick jump* capability on some players is a must for some program applications, especially in the games and entertainment categories. Review the special features of each player and check them against the demands of your program design.
- While some manufacturers say that their prices are fixed, prices can often be negotiated, especially when large quantities or package deals are involved. It is also important to match your needs to the cost of the player. You may not always need the high end product. *Search time* is a feature often traded for cost. For many applications, the search times of intermediate, lower-priced models are more than adequate to do the job.
- Tracking—There is a difference between players and their tolerance for problems with the disc itself. Some will track over warped or damaged areas and still give an acceptable picture, others won't. Some will go crazy when they hit a bad frame on a disc, others won't. Some play the outer edges of discs, others turn frames 49,000 to 52,000 to mashed potatoes.

Unfortunately, there is no great "Consumers' Report" type index of disc players. The best way to work it is to do your own comparative analysis. Put the players side by side, get a few old, beat-up discs of your own and see what happens. If the manufacturer won't allow a side by side comparison, that gives you a quick answer to the question of support. Right?

THE CD REVOLUTION

As noted at the very start of this book, there are a variety of new interactive video formats on the horizon. These formats offer to replace the conventional 12-inch laser videodisc with acceptable video from the surface of a compact disc. Some of these systems may be built into computers; others will be inexpensive computer peripherals. Still others are actually stand-alone "computers in disguise" (they have all the brains of a computer packaged in a box that has everything but a keyboard).

It should be noted that CD-ROM is *not* one of these systems. CD-ROM is a system that uses the CD to store digital data in *read only memory* (ROM). While this data can take many forms, it cannot, as of now, take the form of full-motion video.

The Timetable for Video CDs

Several of the video CD systems are trying desperately to make it into the marketplace by the end of 1988. Some may succeed, although real market penetration probably will not happen until several years after that. Table 11.3 looks at the four most common video CD formats. All are capable of executing the designs presented in this book.

THE DESIGN STATION

The last stop on our tour of computer/video integration is to take a look at future components tied together in a workstation that will help you simulate the interactive video before you've gone through the high cost of video production. Everyone knows that video production costs are high, reshooting is expensive, etc. Even with all the storyboards in the world, it is hard to know how a real interactive video program will work until it is actually in use. The thing that is hard to anticipate is how the interactions will turn out. A computerized system that allows the designer to branch between storyboard frames, either in a digitizer or on tape, may bring the quality of the interactions into sharp focus. A system that mixes rough graphics, character generation, videotape sequences and the control of the computer will give the video producer a way to see how all the elements interact before he/she goes on location and spends tens or hundreds of thousands of dollars producing the video itself.

Figure 11.3 is a hypothetical development system that *can* control a disc player but more important, it can substitute graphics for full production video. In this way the interactions can be looked at, checked out and made perfect before production.

CONCLUSION

In this chapter we have looked at the variety of ways computers work in interactive video systems. We have suggested that, regardless of the job, the computer's function begins with some basic commands that it sends to the disc player. These commands (and the other functions of the computer) are made more accessible when you use an authoring system. We also reviewed some standards for selecting authoring systems and hardware configurations.

Table 11.3:   Video on Compact Disc

	CD-I	CD-V	CVD	DVI
Originator	Phillips/Sony, as part of CD-Interactive specification.	Phillips/Sony, as part of CD-Video specification.	Invented by Lowell Noble and Ed Sandberg. Licensed by SOCS Research, Inc., Los Gatos, CA. (Specifications subject to change.)	Developed by RCA's David Sarnoff Research Center, Princeton, NJ. Basic technology patents retained by General Electric which bought RCA. (Specifications subject to change.)
Licensees	All CD licensees.	All CD licensees.	Mattel, Hewlett-Packard, McDonnell-Douglas, Sanders (Lockheed), Haitai (Korea)	(To be announced)
Encoding	Digital compression—DYUV and run-length encoding. Custom encoder hardware installed in CD-I production studios.	Analog. Uses standard LaserVision production equipment.	Hybrid analog/digital. Uses standard LaserVision analog mastering equipment and SOCS Research custom modulation hardware for additional digital processing.	Digital compression on VAX computer at 30 seconds per frame. Parallel processor, targeted at three seconds per frame, will be online in 1988.
Playing Time, Disc Speed	72 min. partial-screen motion video (plays at normal CD speed, 200-500 rpm)	5 min. CLV full-screen motion video (for NTSC video standard) (plays at up to 2,700 rpm)	20 min. CLV, 10 min. CAV full-screen motion video (plays at 900-1,800 rpm)	72 min. full-screen motion video (plays at normal CD speed, 200-500 rpm)
Picture Resolution/ Quality	384 x 280 full-screen for partial screen motion video. Very high quality.	Excellent-quality LaserVision full-motion video.	High overall quality full motion video. (40 + dB S/N ratio)	256 x 240 for full-screen motion video at 30 frames per second. Comparable to VHS quality.
Video Still-Frame Capability	Up to 5,000 natural pictures per disc (384 x 280 resolution). Very flexible sound-over-still capability.	Not available in consumer player. Up to 9,000 stills when used with frame storer.	Up to 18,000 still frames per side of disc. Very flexible sound-over-still capability in professional unit. Limited sound-over-still capability in Mattel unit.	Up to 7,000 stills at high resolutions (768 x 480). Up to 10,000 stills at medium resolution (512 x 480). Up to 40,000 stills at low resolution (256 x 240). Very flexible sound-over-still capability.
Accompanying Audio Track	Flexible ADPCM fidelity levels.	CD Digital Audio.	CD Digital Audio.	Flexible ADPCM fidelity levels*
Video Decoding Chips	To be manufactured by Matsushita.	Uses established LaserVision decoding electronics.	To be manufactured by Hewlett-Packard.	To be manufactured by GE and future licensees.

Table 11.3:   Video on Compact Disc (Cont.)

	CD-I	CD-V	CVD	DVI
Hardware Configuration	Stand-alone player includes micro-processor, operating system, A/V processors. Available fall 1988 for $1,000–1,500.	Stand-alone player, no interactivity or data storage. Available for about $800.	Mattel stand-alone player, with built-in computer, available Christmas 1988 for $400. Computer peripheral disk drive, with full audio/video/data capabilities available late 1988 for about $1,000.	Set of three add-in boards available in 1988 for $3,000–$5,000. Requires MS-DOS computer and standard CD-ROM drive.
Target Markets	Home entertainment, education, and self-improvement. Also school market for crossover between home and school.	Home entertainment.	Home entertainment and toy market, as well as professional and commercial applications.	Professional and commercial applications, with growth to consumer market.

The ADPCM audio encoding schemes for CDI and DVI are not compatible with each other.)

Source: Reprinted with permission of *CD-ROM Review,* Copyright © 1988 IDG Communications/Peterborough.

Figure 11.3:   Hypothetical Development System

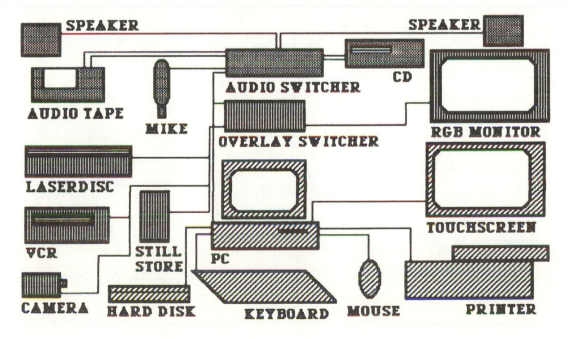

In the final analysis, the evolution of easy-to-use authoring tools and systems has been very long in coming. In fact, some designers say that the definitive, universal authoring system—the one that links almost any computer with almost any disc player, is not yet here. But there certainly are plenty of systems available with which anyone can get started.

The following appendix presents a case study in a simple authoring activity.

APPENDIX: A CASE STUDY IN DISC PROGRAMMING

Let's get a little more specific about what it takes to program a disc by going through a very simple programming assignment. We will not be working with an advanced interactive video design here but one simple enough to be followed by people who, up until now, have relied on professional programmers to execute their designs, or who are about to get into their first authoring language.

For the sake of this example, assume that you are an art teacher who is facing an open-house and who wants to prepare a demo for your guests. You are going to take one of your favorite anthologies of video art (the Voyager disc, *The Dream Machine*) and create a controller that will allow people to access and study four of the best examples of video art on the disc.

You have at your disposal, a Macintosh computer equipped with a hard disc and *Hypercard,* a Pioneer LDV 4200 Laserdisc player and the *Voyager VideoStack,* which is an easy to use, authoring tool.

Hypercard is an authoring system that provides a programming language and interface tools. The video stack provides extended commands (specifically XCMDs) to make *Hypercard* able to talk to the videodisc player. *Hypercard* uses an interpreted language, making it slow for complicated tasks, but, for the tasks in this example, you will not notice any problems with speed. (Note: if you are not familiar with Macintosh operations this description may not be as easy to follow, but you should still be able to recognize the major steps of the authoring task.)

This will be a two-screen system with the disc images showing on your classroom TV set. The Macintosh screen will show the controller what your guests will use to access the sequences.

The Controller

The controller will allow your visitors to select any one of four computer graphic sequences, play the sequence and study it by viewing it in fast or slow motion or by freezing on any frame they like. They can also choose to see the video with straight musical accompaniment or with a narrative explanation. The controller you will create is pictured in Figure 11A.1.

The *Voyager VideoStack* authoring tool comes equipped with it's own set of control buttons. These buttons will allow you to search the disc and find the sequences you like. You can also freeze a frame and display its frame number so you can find the beginning and end frames of all four sequences.

It is necessary to write the beginning and end frame numbers down because you will need them in your programming. Table 11A.1 shows the frame numbers for

Figure 11A.1: The Dream Machine Controller

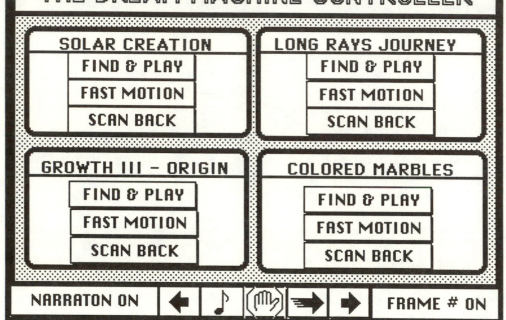

the four sequences we picked for our controller. There is a beginning and end frame for each sequence.

While you are at the *Voyager* authoring tool you will notice that you can access the "scripts" or "sets of commands" for each button. You need to look at the

Table 11A.1: Critical Frames for Programming the Controller

Sequence	Start Frame	End Frame
Creation	15580	16472
Growth III	27400	29350
Long ray's	16512	19216
Colored marbles	19270	20261

Figure 11A.2: New Button Card from the Revised *Voyager* Video Stack

Reprinted with permission of Voyager.

scripts and note how the commands are stated because you will use that exact wording when you prepare your own controller buttons. Figure 11A.2 illustrates the Button Card from the revised *Voyager* video stack. (*Hypercard* offers an easier way to create controls. You can *copy* any button and all its commands. But for the purposes of this lesson we want to look at the command scripts.)

The first thing you need to do to make your controller is add a new card to the *Hypercard* stack. To do this in *Hypercard,* just use the Macintosh mouse to double click on the "new card" command under the *edit* menu.

Next, you create the graphic of the controller by using the paint menu of the Macintosh (something we will not describe here), then you begin adding buttons to your controller.

Hypercard allows you to create new buttons by double clicking on the "new button" command under the *objects* menu. A button appears in the middle of the screen, which you drag with the mouse to the appropriate point. We suggest you position your buttons first and then go in and tell them what to do.

In our *Dream Machine* Controller we have the ability to find and play four different sequences. These sequences can be played faster and we can scan back to the beginning of each sequence at any time. We also have general commands that let us freeze any picture, play it forward or backward in slow motion, and then resume play at normal speed. We have also given the controller the ability to let us turn the frame numbers on or off, and we will be able to choose between two audio tracks— one is music only, the other is a narrative about the sequence.

So, we have positioned the new buttons on the controller, but each button is merely marked, "new button," and does not do anything. We need to go in, name the buttons and "script" their activities. To do this, use the Macintosh mouse to select the button tool from the tool menu and then double click with the mouse on the button to call up the menu for each button.

The menu allows you to select the shape of the button. You can choose an icon or you can type in a name. (In our example, we are mixing button names and icons.) Finally, by double clicking on the script box, you are given access to a card on which you can type the instructions that tell the button what to do.

Assembling Button Commands

Let's say that we are making the controller for the "Creation of the Solar System" segment, created by Optimus. We have already found the frame numbers, and the wording of the standard commands by working with the *Voyager* authoring

Table 11A.2: Wording for the *Dream Machine* Controller

Button	Command Wording
Find & play	video search, 29350
	video play, till 27400
faster	video fast, till, 29350
scan back	video fast, rev, till, 27400
freeze ( )	video step
slo-mo (→)	video slow
slo-reverse (←)	video slow, rev.
resume play (⇒)	video play
music (♪)	video sound, 1, on
narrative	video sound, 2, on

tool. We put the commands for search and play together and come up with instructions for a button we call **Find and Play**. The instructions look like this:

```
on mouse up
    Video search, 27400
    video play, till, 29350
end mouse up
```

The "mouse up" and "end mouse up" wording is provided as a prompt by Hypercard, so our job is to put the key words and frame numbers in between these prompts.

Table 11A.2 shows the commands that go with the buttons on our *Dream Machine* controller.

The most complex command on our controller is the "Frame # on" button because it turns the frame number on and then changes the wording on the button to "frame # off". It is actually a conditional statement. If the frame number is on, then the button says and does one thing. If it is off, then it says and does the opposite. We can create our own button because we've copied the script from the button on the authoring tool. Here's how that command is worded:

```
if the short name of me is "frame # on" then
    Video frame, on
    set the name of me to "frame # off"
else
    video frame, off
    set the name of me to "frame # on"
end if
```

Note the positioning of the commas in each of these statements. They supposedly separate arguments. Without them the command will not be executed.

Now that we have identified the frame numbers and the wording of the commands, created the graphic representation of our controller, and made one button work, the job for the next 20 minutes or so will be to enter in the commands for each of the buttons. It will then be necessary to test each of them to see that they work. The whole process will take about an hour. Not a bad time investment to create a very effective demo tool to use during your open house. (It could work for a trade show exhibit too, of course.)

We hope that this has been a reasonable introduction to the routine of disc programming and to the strings of "if . . . then" commands you will have to put together to bring a complex interactive program to life.

As a point of comparison Figure 11A.3 shows a sample of videodisc control code written in BASIC.

Figure 11A.3: Sample Videodisc Control Code in BASIC.

```
1000 REM intro section
1010 SEGSTART$="20"
1020 SEGEND$="1024"
1030 GOSUB 2000
1040 IF NOT DISCDONE THEN GOSUB 4000      'wait until finished with segment
1050 PRINT "Continue? (y/n) ";
1060 INPUT C$
1070 IF C$="y" OR C$="Y" THEN 1100
1080 PRINT "Bye"
1090 STOP
    .
    .
    .
2000 REM search and then play segment segstart – segend
2010 SECCMD$ = DREPEAT$+SEGEND$+DENTER$+"1"+DENTER$
2020 PRINT #1, DSEARCH$+SEGSTART$+DENTER$;
2030 GOSUB 3000    'wait for search to finish, then send play
2040 RETURN
```

12 Interactive Video: The Outer Limits

To understand the ultimate possibilities of interactive video, we have to understand what humanity is up to in its pursuit of technology. From the broadest perspective, we can imagine that humans are up to nothing short of re-inventing themselves. Robotics often looks like an effort to recreate the body of a human being, eventually without the frailties. But technology, it seems, has chosen to put more of its efforts into recreating the human mind.

We see interactive video as a useful part of this new mind that technology is working so feverishly to create. To understand how interactive video fits into this mind, we have to ask ourselves: what does a mind do and what does a mind need? Three answers to that question seem particularly well suited to interactive video:

1. Storage and retrieval of visual data,
2. Imagination (simulation), and
3. Creativity.

Computers already do all of these things, but interactive video can help computers do them better. Let's take a brief look at the outer limits of each of these applications.

INFORMATION STORAGE AND RETRIEVAL

We know that the computer can store data. With the advent of CD-ROM and other new data storage media, the amount and the quality of the data stored has improved greatly. What CD-ROM and most other media cannot store is what we used to call real images—photographic quality images that can move. This capability, so far, is reserved for video.

The first big jump toward the outer limits of advanced interactive video systems will be mammoth data storage applications and information centers most probably located in schools, libraries and the home. These information centers will have thousands of pages of data that can be easily accessed by artificial intelligence systems, but which can also be linked by the new "free-association" capabilities provided by *Hypercard* and programs like it.

The important thing for interactive video in systems of the future is that the databases will not be limited to just words and graphics. They will contain video and audio as well. So, if you have the world's greatest music information center, and you look up Mozart, you may not only be able to read about him but you will also be able to hear the Jupiter Symphony and see it performed as well. You will also be able to see how Mozart was portrayed in a dozen different movies and you will be able to read the scores of his music. Not only that, if the music database has artificial intelligence, it will tell you: "Based on what I know about you, I think that this is the Mozart piano concerto that you would like best." It doesn't stop here, either; you will be able to go to a section where you can punch in a few notes and the computer will create a Mozart-like piece based on those notes. It will then explain what makes the piece "Mozartian."

The possible applications for information systems such as this in areas like health education, geography and astronomy, not to mention general data storage systems like encyclopedias and atlases, are great. But let's stay with our music analogy and take it one step further. What if the system not only gave you data about Mozart, but let you talk to him?

We are now talking about the next level of future activity to expect from interactive video—simulation.

MEGA SIMULATION SYSTEMS

As we said earlier, it seems that technology is moving toward the re-creation of the human mind. Consider the simulation systems of the future as the imagination of this computer mind we are creating. To figure out the best applications of the new technology we have to figure out the favorite applications of our own imaginations.

Think of a system built around the familiar experience of an automobile trip. With computer overlay you can create the instrument panel of any kind of car you want, and with a few extra buttons, you can create a vehicle for taking simulated trips with all kinds of extra benefits thrown in.

A mega simulation system for travel could allow you to drive through the national parks of any country on earth. There could also be a road atlas to display the maps of the areas you are visiting. For added adventure on these imaginary trips, you could go "off road" for something exciting that might resemble a fast-paced arcade game, rather than a simple Sunday drive. For example, you could tool up a winding mountain road or charge across the wild terrain of some desert badland.

Simulated trips such as these also have educational potential. We have already mentioned the maps that can be stored in the database, but these mega simulations of driving can also act as driving simulators that contain driver's education manuals for several states.

A superb car stereo and a communications system, based on a CB radio model, could also be added to your simulator as well as a car telephone and a small TV screen so a section of your windshield could show you the person you are talking to on the CB radio or telephone.

If you get into mechanics, you could learn car repair techniques and perhaps get your simulator to the point where you could enter a simulated Indianapolis 500.

Eventually, these simulators could become fully enclosed, hydraulically operated systems that would make you feel like you were behind the wheel of an actual vehicle. They might include pedals and a gear shift, and could pitch and lurch as you travel. They might also include extra seating so you could take a friend along for the ride. To use the ultimate extent of your imagination, you could turn your simulated vehicle into a time machine or a spaceship.

The mega simulation systems of the future could become fantastic entertainment devices. Their creation would be a blend of cabinet-making, hydraulics, computer programming, video production, graphics overlay and database assembly.

CREATIVITY

The outer limits of interactive video could make it part of an instrument to enhance creativity as well as allow for simulation and information management.

Interactive Art

Art can be representational or nonrepresentational. In the representational vein, we talk about simulations of situations in which the viewer can become more than a viewer—he or she can become a participant. The user of an interactive drama can at least effect the outcome of the story. But, in a more imaginative application, he or she can *participate* in the story, become one of the characters and experience a different time and place, or a new situation. The user can live through danger, adventure or romance.

In a nonrepresentational video, the user can still control movement through space and time, and can control sound. Imagine something like the Windom Hill

Series of video music discs (one of the discs shows beautiful images of autumn leaves accompanied by music). Put images such as these on a CAV instead of a CLV disc, so that you have frame-by-frame control backward and forward. Now mix abstract impressions with the representational images and disassociate the music from the picture so it will play whether the images are going backward or forward. This can be done with a separate source of audio such as a CD, but a videodisc soundtrack should also be maintained with cues that are in sync with the picture. Put in a control so you can mix the soundtracks. Add a computer graphic overlay system so the viewer can add images on top of the mix of representational images and abstract impressions that are on the videodisc. An audio record capability could also be added so the user can add new audio. What have you got? Something like a "do-it-yourself music video kit" or a "do-it-yourself video art kit."

Interactive Video as a Medium for the Creation of Art

What would the do-it-yourself interactive video art kit do for the user? Well, you could create your own audio/video entertainments; you could make your own rock videos; you could also create your own mood videos (classical, jazz, rock or whatever). Then you could impress your friends, entertain them or bore them silly with your artistic creations. You could splice up your home movies, edit them, and add titles or special effects. These could be intercut with premade "clip art" or movie clip sequences.

Let people play with the players. Mel Brooks and his troupe could give us a great low-brow do-it-yourself interactive video comedy disc. Woody Allen with his understanding of visual and audible humor and the way the two mix would give us more sophisticated material. We can create interactive video images to the accompaniment of music—some representational, some not. We can also convey moods—excitement, sorrow, joy, humor, peace and intense anger. Now the question is, what would be the effect of these creations outside the realm of entertainment?

Interactive Video and Psychology

On the positive side, an interactive video room that projects peaceful images on the walls to the accompaniment of peaceful music should have a restorative effect on the psyche of just about anyone. The ability to effect the direction and general sense of the music and the images might also be helpful.

Media rooms are not a new idea, neither are interactive media rooms. Some great work in the area of interactive media rooms has been done at MIT, but the impact of the psychological effects of an interactive video media room and the isolation of *interactivity* as a major variable in the effectiveness of the room needs to be studied further. It will surely yield some very interesting results.

CONCLUSION

In the early 1980s, promoters of interactive video claimed that the technology had great potential. What has happened since, however, suggests that the success of interactive video may have been greatly overestimated.

The benefits that we (the designers of interactive video systems) lauded so greatly just have not impressed the general public enough to make any single application of interactive video the "breakthrough application"—the breakthrough that drives people into the stores demanding interactive video systems. No application has made corporate chairpeople stand up and say, "If there's one thing I know, it's that this company needs an interactive video network."

What this book has tried to suggest is that the slow growth of interactive video has been the result of a specific cause—it is difficult to do interactive video well. It is so difficult, in fact, that most attempts at producing interactive video programming don't even look like failed attempts at something that could be good; they look like successful attempts at something that could never be good. The result is that potential customers say "so what!"

The solution is obvious—people have to stop making bad interactive video. They have to concentrate their energies on a very high level of quality. They have to strive to make and show only successful examples of interactivity. If their best efforts do not work out they have to scrap them before they see the light of day. At the same time, the kind of competition and secrecy that usually surrounds new products has to be done away with. People have to share the whole technology or it will die.

Interactivity is more than just a new trick, a new device or even a new technology. Interactivity between user and medium—real two-way communication—is something totally new and very different. The fact that a lot of people saw interactivity as a way to make a quick buck is too bad. It doesn't invalidate the technology, it just means that good interactive video will have to be pursued and developed by people with a little more patience and altruism. These people have to understand that they are uncovering something that could lead to quantum leaps in the way people learn and work. It could also add a new dimension to art and entertainment. At the very least, interactive video should *humanize* the computer so that anyone, anywhere, can turn on a computer and use it immediately, whether they are familiar with computers or not. The road to such a futuristic computer system starts with on-line-video-help and ends with a computer that talks to you, answers your phone, takes your dictation, calls up data when you ask it to, makes your airline reservations, updates your AV presentations, samples your TV shows, tells you which tie goes with which suit, and whether or not you should bother to read the number one best-seller.

Moreover, you should be able to *pick* the computer's on-screen personality according to his or her looks, voice, interests, experience, etc. Picking a computer's personality will be like hiring an administrative assistant. And all the time that computer should have instant access to unlimited stacks of data in whatever form is needed, including full-motion color video.

Signs of Success

If we could look for one sign that interactive video technology is going in the right direction, it would be that within five years, instant-on-line-video-help would be a reality that would enable everyone to use any computer and any software immediately, without having to read a manual or take a course. The futuristic computer described above is a possible descendant of such a system.

Perhaps the most impressive product of such technology was described by Doctor Heimlich (inventor of the Heimlich maneuver and an ardent advocate of "Computers for Peace"). Dr. Heimlich believes that improved understanding between all of humanity can be the ultimate outcome of the communications revolution, and may also be the best hope for world peace.

Finally, if the applications we have described do not impress you as being the true outer limits of interactive technology, we suggest that you look into the work being done at MIT's Media Lab and read Stewart Brand's book, *The Media Lab: Inventing the Future at MIT* (see Bibliography at the end of this book). At MIT they are trying to do more than re-invent the human mind. According to Mr. Brand, they are trying to "re-invent the world."

Bibliography

Ambron, Sueann and Hooper, Kristina, eds., *Interactive Multimedia,* Redmond, WA: Microsoft Press, 1988.

Bennion, Junius L., "Authoring Interactive Videodisc Courseware." In *Videodisc-Microcomputer Courseware Design,* edited by Michael L. DeBloois. Englewood Cliffs, NJ: Educational Technology Publications, 1982.

Brand, Stewart, *The Media Lab, Inventing the Future at MIT,* New York: Viking, 1987.

Burke, James, *The Day the Universe Changed,* Boston, MA: Little Brown & Co., 1986.

CD-ROM, The New Papyrus, Redmond, WA: Microsoft Press, 1986.

Daynes, Rod, ed., *The Videodisc Book,* New York: John Wiley & Sons, 1984.

Dunton, Mark, and Owen, David, *The Complete Home Video Handbook.* New York: Random House, Inc., 1982.

Eisenstein, Sergei, *Film Form,* New York: Harcourt Brace Jovanovich, 1969.

Flemming, Malcolm, and Levie, W. Howard, *Instructional Methods Design.* Englewood Cliffs, NJ: Educational Technology Publications, 1978.

Floyd, Steve, and Floyd, Beth, *Handbook of Interactive Video,* White Plains, NY: Knowledge Industry Publications, Inc., 1982.

Gagne, R.M., and Briggs, L. J., *Principles of Instructional Design,* 2nd ed., New York: Holt, Rinehart and Winston, 1979.

Goodman, Danny, *The Complete Hypercard Manual,* New York: Bantam, 1987.

Iuppa, Nicholas V., *A Practical Guide to Interactive Video Design,* White Plains, NY: Knowledge Industry Publications, 1984.

Joels, Kerey Mark, *The Mars One Crew Manual,* New York: Ballantine Books, 1985.

Lambert, Steve and Sallis, Jane, *CD-1 and Interactive Videodisc Technology,* Indianapolis, IN: Howard W. Sam & Co., 1987.

Langdon, Danny G., *Instructional Designs for Individualized Learning,* Englewood Cliffs, NJ: Educational Technology Publications, 1973.

Lee, Robert, and Misiorowski, Robert. *Script Models,* New York: Hastings House, 1978.

Martin, James, *Viewdata and the Information Society,* Englewood Cliffs, NJ: Prentice-Hall, 1982.

Schwartz, Ed, *The Educators' Handbook to Interactive Videodiscs,* Assoc. for Educational Communications and Technology, 1985.

Videodisc Monitor, editors, *Videodisc and Related Technologies: A Glossary of Terms,* Future Systems, Inc., 1986.

Index

ABOUT THE AUTHORS

Nicholas V. Iuppa is an experienced media manager as well as an active writer and producer of interactive videodiscs. He is currently serving as Product Development Manager, Consumer-New Technology for Apple Computer, Inc.

Previous to this position, Mr. Iuppa was Managing Director of Interactive Video for Hewlett-Packard's Television Network where he worked on innovative applications of videodisc technology.

Prior to his position at Hewlett-Packard, Mr. Iuppa was vice-president of media production at ByVideo Inc., where he and his staff produced more than a dozen interactive videodiscs for the ByVideo shopping system. Their work was recently recognized at a National Videodisc Awards Presentation at the University of Nebraska. There, the ByVideo system was honored as the Best Overall Videodisc Project of the year.

Previously, as Vice President, Director of Media Services at Bank of America in San Francisco, Mr. Iuppa supervised a large-scale media center where his staff produced over five hundred video programs, including several of the very first examples of complex interactive video training.

Mr. Iuppa's writing credits include story development assignments at MGM and Walt Disney studios and instructional design work at Bank of America and Eastman Kodak. He is also the author of technical articles, popular magazine fiction, a technical work titled *A Practical Guide to Interactive Video Design*, and a humorous compendium of management tactics titled *Management by Guilt*.

Karl Anderson is a consultant in interactive video design and was software engineer and designer of interactive videodisc games at ATARI.

145